# Lithium-Ion Battery Failures in Consumer Electronics

For a listing of recent titles in the
*Artech House Power Engineering Library,*
turn to the back of this book.

# Lithium-Ion Battery Failures in Consumer Electronics

Ashish Arora
Sneha Arun Lele
Noshirwan Medora
Shukri Souri

**ARTECH**
**HOUSE**
BOSTON | LONDON
artechhouse.com

**Library of Congress Cataloging-in-Publication Data**
A catalog record for this book is available from the U.S. Library of Congress.

**British Library Cataloguing in Publication Data**
A catalog record for this book is available from the British Library.

ISBN-13: 978-1-63081-603-2

**Cover design by John Gomes**

**© 2019 Artech House**
**685 Canton Street**
**Norwood, MA 02062**

All rights reserved. Printed and bound in the United States of America. No part of this book may be reproduced or utilized in any form or by any means, electronic or mechanical, including photocopying, recording, or by any information storage and retrieval system, without permission in writing from the publisher.

All terms mentioned in this book that are known to be trademarks or service marks have been appropriately capitalized. Artech House cannot attest to the accuracy of this information. Use of a term in this book should not be regarded as affecting the validity of any trademark or service mark.

10 9 8 7 6 5 4 3 2 1

# Contents

Preface  13

# 1
## Lithium-Ion Cells  17

1.1 Other Cell Chemistries  18
    1.1.1 Lead Acid  19
    1.1.2 NiCd  19
    1.1.3 NiMH  19

1.2 Li-Ion Cell Operation  20

1.3 Li-Ion Cell Varieties  21
    1.3.1 Cathode Materials  22
    1.3.2 Anode Materials  23
    1.3.3 Separator  24
    1.3.4 Electrolytes  25
    1.3.5 Current Collectors  25
    1.3.6 Solid Electrolyte Interface Layer  25

1.4 Cell Construction  26
    1.4.1 Prismatic Cells  26
    1.4.2 Cylindrical Cells  26

1.4.3 Pouch Cells 27

1.5 Battery Management System for Li-Ion Cells 28

1.6 Charger Circuits for Li-Ion Cells 28

1.7 End of Life 29

1.8 Shelf Life 31

1.8.1 Life Test and Verification Process 32

1.9 Cycle Life 33

References 34

# 2

## Overview of Li-Ion Battery Systems 37

2.1 Single-Cell Li-Ion Battery Systems 38

2.2 Multicell Li-Ion Battery Systems 41

2.2.1 Other Configurations 43

2.3 Automotive Battery Systems 43

2.4 Battery Systems for Large-Scale Energy Storage Systems 44

References 46

# 3

## Power Supplies for Portable Consumer Electronic Products 47

3.1 Components of a Typical AC-DC Converter (AC Adapter) Circuit 48

3.2 AC Adapter Requirements 52

3.2.1 Propagating Circuit Board Failures 53

3.2.2 Burn Hazards 54

- 3.2.3 Component Derating 60
- 3.2.4 Power Supply Efficiency 63
- 3.2.5 Single Points of Failure 64
- 3.2.6 Power Supply Connectors 66
- 3.3 Adapter Specifications 67
- 3.4 Power Supply Construction and Assembly Issues 67
  - 3.4.1 Input Power Connection 68
  - 3.4.2 Electrolytic Capacitors 69
  - 3.4.3 Mechanical Damage and Access to High Voltages 71
  - 3.4.4 Creepage and Clearance Distances 71
  - 3.4.5 Soldering/Contaminants and Other Assembly Issues 73
- 3.5 Testing AC Adapters 74
  - 3.5.1 Electrical Characterization Tests 74
  - 3.5.2 Thermal Characterization Tests 75
  - 3.5.3 Mechanical Abuse Tests 76
  - 3.5.4 Single Points of Failure Tests 76
- 3.6 DC-DC Converter Circuits 77
- References 79

# 4

# Li-Ion Battery Pack Charge Circuits 83

- 4.1 Charge Profile 84
- 4.2 Safety Timers 86
- 4.3 Battery Temperature Monitoring 87
  - 4.3.1 Charging Li-Ion Cells at Low Temperatures 87
  - 4.3.2 Charging Li-Ion Cells at High Temperatures 89
- 4.4 Battery ID 89
- 4.5 Charger Specifications and Requirements 90

4.6 Charger Circuit Construction and Assembly Issues 92
4.7 Testing Charger Circuits 93
4.8 Wireless Charger Circuits 98
References 100

# 5

## Battery Protection Circuit Consideration 101

5.1 Need for a Protection Circuit 101
5.2 Single-Cell Battery Packs 106
5.3 Multicell Battery Packs 111
5.4 Large Battery Packs 115
5.5 Battery and PCM Specifications 116
5.6 PCM Design Review Tools 120
5.7 PCM Construction and Assembly Issues 121
5.7.1 Single-Cell Battery Packs 121
5.7.2 Multicell Battery Packs 122
5.8 Cell Failure Predictions 124
5.8.1 Cell Resistance as a Predictor of Failure 125
5.9 Testing PCMs 127
5.10 Summary 130
Appendix 5A: Electric Shock Hazards 130
References 134

# 6

## Industry Standards and Testing 135

6.1 Commonly Used Standards to

Evaluate Li-Ion Cells and Batteries in Portable Consumer Electronic Devices 135
   6.1.1 UN Transportation Requirements 136
   6.1.2 Underwriters Laboratories Testing Requirements 137
   6.1.3 IEC Standards 143
   6.1.4 IEEE Standards 145
   6.1.5 Comparing Standards 145
6.2 Other Tests 146
   6.2.1 Forced Internal Short-Circuit Test 146
   6.2.2 Nail Penetration Test 149
   6.2.3 Thermal Stability Test 150
   6.2.4 Dent/Pinch Test 151
6.3 Predicting Li-ion Cell and Battery Shelf Life 151
   6.3.1 Capacity Degradation 152
   6.3.2 Calendar Aging 153
   6.3.3 Fit Method 153
   6.3.4 Time Dependence 154
   6.3.5 Summary 155

Appendix 6A: International Organizations and Standards 155

Appendix 6B: Common Tests in Industry Standards 157

References 160

# 7

# Physical Construction of Battery Packs 163

7.1 Single-Cell Battery Packs 164
   7.1.1 Soldering of Cell Tabs 164
   7.1.2 Routing of Battery Pack Wires 166
   7.1.3 Cell Tab Insulation 167
   7.1.4 Circuit Board Insulation and Mounting 168

7.1.5 Contaminants   169
7.2  Multicell Battery Packs   173
   7.2.1  Routing of Voltage Sense Wires   173
   7.2.2  Separation and Insulation of Solder Joints   174
   7.2.3  Tab Placement and Spot Welding   174
7.3  Larger Battery Packs   175
   7.3.1  Excessive Length of Voltage Sense Wires   175
   7.3.2  Improper Wire Routing   176
   7.3.3  Inadequate Insulation   177
   7.3.4  Improper Torqueing of Screws   178
   7.3.5  Cell Separation   179
   References   180

# 8

## Field Failures and Investigation Tools   181

8.1  The Scientific Method for Investigating Battery Failures   182
   8.1.1  The Scientific Method   182
   8.1.2  Applying the Scientific Method to Battery Failure Investigations   183
8.2  Analyzing Battery Failures   185
8.3  Battery Failure Root Cause Analysis: A Case Study   188
   8.3.1  Background   188
   8.3.2  Battery System Design   189
   8.3.3  Visual Inspection and X-Ray Analysis   190
   8.3.4  Battery Charger Circuit Review and Evaluation   191
   8.3.5  Battery Protection Circuit Review and Evaluation   191
   8.3.6  Likely Cause of Failure   192

8.3.7 Cell Construction Review   195
8.3.8 Summary   197
Appendix 8A   Investigating Failures   197
Cause of Failures in Electrical Equipment   198
Overview of Tools Used when Conducting an Analysis of the Cause of the Failure   198
References   203

# 9

## Checklists   205

9.1 Charger Checklist   206
9.2 Battery Checklist   206

Glossary   215
About the Authors   225
Index   227

# Preface

Lithium-ion (Li-ion) batteries power hundreds of millions of devices in the field today. These batteries are used in a wide variety of applications ranging from toys, smartphones, cars, and even airplanes and ships. The advantages and disadvantages of the Li-ion chemistry over other competing battery chemistries are well documented in the literature. Li-ion cells typically have a restricted range of operating voltages, currents, and temperatures. Operating the cells outside their specifications increases the risk of a battery failure in the field. For this reason, most Li-ion battery systems utilize sophisticated charging and protection circuitry to prevent the cells from operating outside these specifications.

The design of the charging and protection circuitry varies from one application to another. Li-ion batteries used in automotive applications have different design and performance requirements as compared to Li-ion batteries used in portable consumer electronic devices. The aim of this book is to introduce the reader to Li-ion battery systems, especially those used in portable consumer electronic devices. This book does not focus on the Li-ion chemistry itself or on how a Li-ion cell operates. The focus of this book is on the integration of Li-ion cells into battery systems in portable consumer electronic devices. The book targets readers from diverse backgrounds who would like to get not only an introduction to Li-ion battery systems in general but an understanding and a working knowledge of how to safely use Li-ion batteries in portable consumer electronic devices. Each chapter in the book attempts to take a hands-on approach based on the years of experience that the authors have accumulated in this discipline, and provides real-world examples of the common pitfalls that may be encountered when using these batteries in portable consumer electronic devices.

Chapter 1 focuses on introducing the reader to the family of Li-ion cells, what the cells look like, how they age in the field and on the shelf, and what to expect in terms of cell performance when using them in portable consumer electronic devices. Chapter 2 provides a high-level overview of the various subsystems of a Li-ion battery system in different applications.

Chapters 3, 4, and 5 delve into the various requirements of a Li-ion battery system used in portable consumer electronic devices. This includes the design of AC adapters and charge and protection circuits used in Li-ion battery systems to ensure that the Li-ion cells do not operate outside their rated specifications. These chapters aim to educate the reader on the design, construction, and testing requirements for these subsystems.

Chapter 6 introduces the reader to the world of certifications and testing requirements for Li-ion battery systems. The focus of this chapter is on standards and testing requirements for portable consumer electronic devices. Chapters 7 and 8 provide practical examples of things that can go wrong when using Li-ion batteries and a methodology to investigate Li-ion battery failures in the field. A case study is presented in Chapter 8 that demonstrates how the methodology can be used in determining the root cause of failure. Chapter 9 provides the reader with checklists that can be used to evaluate a Li-ion battery system or to compare Li-ion battery systems from different suppliers.

We are deeply indebted to our colleague Dr. Quinn Horn who helped with a review of the book. We would also like to thank the team at Artech for their constant support. Last but not least, we would like to acknowledge the unrelenting support from our families. Ashish dedicates this book to his wife Dipali, daughter Anya, and parents Surender and Neelima. Sneha dedicates this book to her husband Satish and mother Vasudha. Noshirwan dedicates this book to Christ Jesus, the God of Heaven who has carried him, comforted him, guided him, and forgiven him. Shukri dedicates this book to his wife Helga, daughters Maysa and Lara, and parents Jeries and Aida, as well as to all striving toward cleaner and safer sources of energy.

# 1

# Lithium-Ion Cells

The term lithium-ion (Li-ion) refers to a family of battery chemistries. The aim of this chapter is to introduce the reader to the Li-ion chemistry and provide a high-level overview of the operation and construction of Li-ion cells. At a high level, a Li-ion cell is a cell in which the negative electrode (anode) and the positive electrode (cathode) materials serve as a host for lithium ions (the term anode and cathode will be used to refer to the negative electrode and positive electrode, respectively, in this chapter and in the remainder of the book). The many advantages of Li-ion cells are well documented in the literature and billions of these cells are used in a wide variety of applications today, ranging from portable consumer electronic devices to electric vehicles to large-scale, grid-based energy storage systems. Table 1.1 [1] demonstrates the reason that most portable consumer electronic devices use Li-ion cells for energy storage and not one of the other competing cell chemistries. Compared to the other competing battery chemistries (e.g., lead acid, nickel metal hydride (NiMh), nickel cadmium (NiCd)), Li-ion cells have a su-

**Table 1.1**
Comparison of Different Battery Technologies

| Property | Unit of Measurement | Lead Acid | NiMH | Lithium-Ion |
|---|---|---|---|---|
| Cell voltage | Volts | 2 | 1.2 | 3.2–3.6 |
| Energy density | Wh/Kg | 30–40 | 50–80 | 100–500 |
| Power density | W/Kg | 100–200 | 100–500 | 500–8000 |
| Useful capacity | Depth of discharge % | 50 | 50–80 | 100–200 |
| Cycle life | Number of cycles | 600–900 | >1000 | >2000 |

Adapted from [1].

perior energy density and can store a greater amount of energy in a smaller package, making them ideally suited for portable consumer electronic devices.

Other fundamental advantages of Li-ion cells include the reduction potential of lithium. Lithium has the lowest reduction potential of any element (–3.04V), allowing lithium-based cells to have one of the highest cell potential.[1] Typical Li-ion cells used in portable consumer electronic devices have a nominal voltage rating of 3.7V, which is comparatively much higher than the 1.2V nominal voltage rating of both NiMH and NiCd cells. In addition, lithium is the third lightest element and has one of the smallest ionic radii of any single-charged ion, allowing lithium-based cells to have a comparatively higher gravimetric and volumetric capacity and energy density [2].

## 1.1 Other Cell Chemistries

Although Li-ion cells are currently the undisputed choice for energy storage in portable consumer electronic devices, several other battery chemistries are still popular in other applications. As an example, lead acid batteries are widely used for vehicle startup and ignition as well as for energy storage in power backup applications

---

1. In theory, the highest potential for a single elemental cell would be achieved using lithium (standard electrode (reduction) potentials in aqueous solution at 25°C = –3.04V) and fluorine (standard electrode (reduction) potential in aqueous solution at 25°C = +2.87), resulting in a single cell with an open circuit voltage of 5.91V. However, due to the extremely active nature of the two elements, as a practical matter such a cell is currently not available

such as in uninterruptible power supply (UPS) systems. NiMH batteries are also extensively used in some consumer electronics as well as in automotive applications. The different cell chemistries have their own unique advantages and disadvantages. Some of the common types are briefly discussed next.

### 1.1.1 Lead Acid

Lead acid batteries consist of a lead dioxide cathode, a lead anode, and a sulfuric acid solution that acts as the electrolyte. Lead acid batteries have a number of different degradation modes that can affect power and energy depending on the exact type of lead acid battery. As an example, for flooded lead acid batteries, as the battery is cycled, the active material dislodges and separates from the plates and cannot be restored. This limits the life of the battery and entails periodic maintenance. Lead batteries tend to be bulkier, especially when compared to the other battery chemistries. However, the technology is mature and reliable, making these batteries popular in industrial applications.

### 1.1.2 NiCd

NiCd cells use nickel hydroxide as a cathode, cadmium as an anode, and potassium hydroxide as the electrolyte. Compared to lead acid batteries, NiCd cells have a longer life and require little maintenance. However, NiCd cells are more expensive than lead acid batteries and have a comparatively higher self-discharge rate and a lower cell voltage. They also have a relatively low energy density, are susceptible to a memory effect, and need periodic full discharges. NiCd batteries are commonly found in applications such as power tools and emergency lighting. However, Li-ion batteries have started to displace these batteries in these applications. In addition, the toxicity of cadmium has resulted in NiCd batteries being banned in the European Union (EU) for most applications.

### 1.1.3 NiMH

NiMH cells use nickel hydroxide as the cathode. Hydrogen is used as the active element in a hydrogen absorbing anode. The electrolyte is alkaline, usually potassium hydroxide. NiMH cells were a popular battery choice for laptops and other consumer electronic applications in the 1990s and early 2000s. Li-ion cells have replaced NiMH cells in most of these applications today. NiMH cells are currently still used in hybrid electric vehicles but Li-ion cells are rapidly becoming the battery of choice in these applications.

## 1.2 Li-Ion Cell Operation

Figure 1.1 shows a schematic of a Li-ion cell. In a Li-ion cell, alternating layers of anode and cathode materials are separated by a porous film (separator). An electrolyte provides the media for the transport of lithium ions. The most common varieties of Li-ion cells use a graphite-based negative electrode and a lithium metal oxide based positive electrode. The cells also consist of an electrolyte that in a Li-ion cell is an organic solvent with a dissolved lithium salt [3].

The active materials in a Li-ion cell operate in an intercalation process where the lithium ions are reversibly inserted and removed

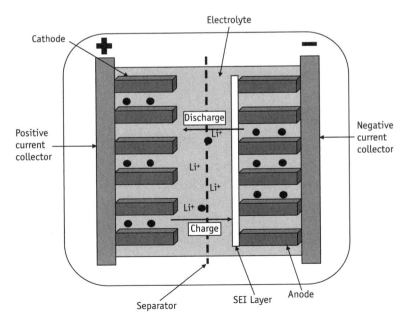

**Figure 1.1** Li-ion cell components.

from the electrodes during operation while no significant change occurs to the structure of the host (i.e., the electrodes during operation). Lithium ions move from the positive to the negative electrode when a Li-ion cell is being charged. The transfer is reversed as the cell is discharged.

## 1.3 Li-Ion Cell Varieties

The four primary functional components of a Li-ion cell are the negative electrode (anode), positive electrode (cathode), separator, and electrolyte. To increase the Li-ion cell's power capability, it is desirable for the anode and cathode materials to have large geometric electrode areas with high porosity to increase the reaction area. For this reason, Li-ion cell electrodes are constructed of a paste composed of fine particles coated on thin current collectors (usually thin copper or aluminum foils) [4].

The choice of electrode materials within the cell and the selection of the electrode/electrolyte combination has been a topic of intensive research and development with Li-ion cells. This choice (often referred to as the Li-ion cell's chemistry), largely differs based on the requirements of the application and has a significant impact on capacity, life, and safety. Table 1.2 [5] lists some of the different materials used for the positive electrode (cathode), the negative elec-

Table 1.2
Li-ion Chemistry Combinations

| Anodes | Electrolytes | Cathodes |
|---|---|---|
| Metallic lithium | Liquid organic electrolytes | $Li_{1-x}Ni_{1-y-z}Co_yM_zO_4$ (M=Mg, Al, etc.) |
| Lithium alloys | Solid polymer electrolytes | $Li_{1-x}Co_{1-y}M_yO_2$ |
| Graphites | Polymer gel electrolytes | $Li_{1-x}Mn_{2-y}M_yO_4$ |
| Other lithiated carbons | Ionic liquids | Polyanionic compounds ($Li_xFePO_4$) |
| Tin-based composite alloys | — | $Li_{1-x}Mn_{1-y}M_yO_2$ (M=Cr, Co, etc.) |
| 3-D metal oxides and nitrides based alloys | — | |

trode (anode), and the electrolyte in Li-ion cells (several materials listed have not made it into commercial cells).

### 1.3.1 Cathode Materials

The three common classes of cathode materials used in Li-ion cells include layered transition-metal oxides, spinels, and olivines. Lithium cobalt dioxide (LCO) is the most common layered transition metal oxide cathode used in Li-ion cells due to its high specific capacity, high volumetric capacity, high discharge voltage, and good cycling performance. One downside of the layered transition-metal oxide cathode material when compared to some of the other alternatives is its relatively lower thermal stability that results in the material releasing oxygen once it is heated above a certain temperature.

As an example, Figure 1.2 [6] shows the surface temperature of a Li-ion cell with a LCO cathode during a test based on a heating test per the Underwriters Laboratories (UL) UL1642 standard where the cell was heated to a temperature of 150°C and soaked at this temperature for 10 minutes. During the test, exothermic reactions within the cell caused its temperature to increase above the

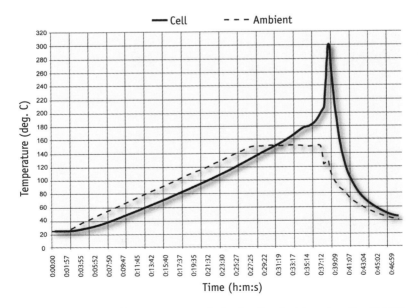

**Figure 1.2** Cell surface temperature during heating test based on the UL1642 heating test.

ambient temperature, eventually resulting in the cell going into thermal runaway, venting, and ejecting its internal contents.

Lithium manganese oxide (LMO) is an example of a spinel type cathode material used in Li-ion cells. While these materials are considered to be comparatively safer than the layered transition-metal oxide type cathode materials commonly used in Li-ion cells, they have a comparatively lower specific capacity and also typically offer lower cycle life when compared to the LCO type materials. Lithium iron phosphate (LFP) is the most common type of material used in the olivine class of Li-ion cells. While cells utilizing the olivine class of cathodes are also comparatively safer than other types of Li-ion cells based on the layer transition-metal class of cathodes, they have a lower nominal cell voltage and also a lower specific capacity that limits their appeal in portable consumer electronic devices.

Extensive research is also ongoing into what can be termed as hybrid cathode materials. As an example, one cathode material that has been extensively researched is the $Li(Ni_{0.5}Mn_{0.5})O_2$ cathode material. The addition of the cobalt into the LNMO has been found to be an effective way to enhance structural stability. Other cathode compounds being researched for use in Li-ion cells include compounds based on sulfur and lithium sulfides, selenium and tellurium, and iodine [7]. Figure 1.3 demonstrates the results obtained by accelerated rate calorimetry (ARC) for different delithiated positive electrode materials (i.e., cathodes) [6].

### 1.3.2 Anode Materials

Carbon-based materials (graphite and hard carbon) are the most common anode materials used in commercial Li-ion cells today. The cathode material used in Li-ion cells is typically the capacity limiting electrode. As an example, the average capacity of LCO type cathode material is approximately 140 mAh/g while the average capacity of commercial graphite is in the range of 330 mAh/g [8]. For this reason there is comparatively more research into suitable cathode materials for Li-ion cells as compared to anode materials for use in these cells.

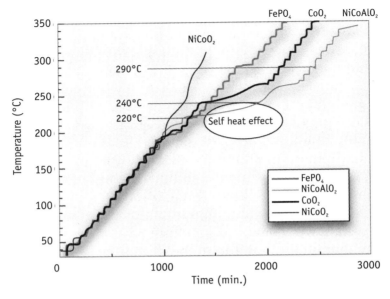

**Figure 1.3** Results obtained by ARC for different delithiated positive electrode materials using liquid electrolyte at the charge state.

### 1.3.3 Separator

The separator in a Li-ion cell plays a critical role. It acts as a physical barrier between the anode and the cathode, preventing an internal short circuit. It is permeable for the lithium ions. Commercially available Li-ion cells use polyethylene- and polypropylene-based separators because of their good mechanical and chemical stability. Separators in Li-ion cells play an additional safety role. The separators of many high-energy cells are typically designed to shut down ionic conductivity in the cell if the cell temperature rises above a certain level. IEEE Std. 1725-2011 requires that separators in Li-ion cells maintain their isolation property for cell temperatures up to 150°C during a temperature excursion for at least 10 minutes to maintain safety of the cell (with the cell at a state of charge exceeding 80%). The standard also requires that the separator exhibit a "rapid internal resistance increase (ionic conductivity shutdown) when cell temperature rises above the shutdown point." This internal resistance increase should be at a minimum of two orders of magnitude at a minimum speed of 2,000 $\Omega.cm^2/s$ [9].

### 1.3.4 Electrolytes

Electrolytes used in Li-ion cells are commonly a mixture of organic solvents such as ethylene carbonate, dimethyl carbonate, and propylene carbonate, which contain a dissolved lithium salt (lithium hexafluorophosphate ($LiPF_6$)). It is common for cell manufacturers to include low concentration of additives that assist in improving cell performance.

### 1.3.5 Current Collectors

The current collector helps in the even transfer of current to the active material in the cells. The current collector also provides mechanical support to the active material and provides a point of mechanical connection to leads that transfer current into the cell. Foils of copper (used as a substrate for the active material for the anode) and aluminum (used as a substrate for the active material for the cathode) are the most common current collectors used in commercial Li-ion cells, especially cells used in portable consumer electronic applications.

### 1.3.6 Solid Electrolyte Interface Layer

As shown in Figure 1.1, the passivating solid electrolyte interface (SEI) layer is critical for the stable operation of a Li-ion cell that uses a carbon-based anode. At typical cell voltages, mixtures of lithiated carbon and the organic electrolyte are not thermodynamically stable, resulting in a reaction between the two. The result of this reaction is the formation of a passivating layer (called the solid electrolyte interface layer) on the carbon surface. The SEI layer is typically formed during the first few charge cycles after a cell is assembled (this occurs at the cell manufacturing facility and is called cell formation). The formation of this layer results in an irreversible capacity loss due to the consumption of charge-carrying lithium ions. Other consequences of the SEI layer during the life of the cell include a gradual loss of cycleable lithium causing irreversible capacity loss, growth of impedance film layer on the negative surface causing irreversible power loss, as well as partly reversible self-discharge in a cell [10].

## 1.4 Cell Construction

Li-ion cells come in a variety of different form factors. Prismatic, cylindrical, and pouch cells are the common configurations of Li-ion cells used in portable consumer electronic devices. Cylindrical and prismatic cells of smaller form factor and capacity (< 4 Ah) are generally wound while larger prismatic cells typically have a flat-plate or stacked construction [3]. Pouch cells may also have a wound structure with conductive foil tabs sealed in a pouch pack.

### 1.4.1 Prismatic Cells

Prismatic cells typically have a metallic rectangular-shaped enclosure traditionally made of aluminum or steel. The metal can is robust and serves as a good heat dissipater. The enclosure is usually sealed via a welding process and allows for a venting mechanism. These cells are available in different sizes and the type of vent and terminal placement may vary [11]. Figure 1.4 shows the top section of a commercially available prismatic cell.

Prismatic cells have a higher energy density due to their compact form factor and are typically available in sizes up to 100 Ah. They are commonly used in portable devices. Large-format prismatic cells are more commonly used in high-power applications such as traction batteries for cars.

### 1.4.2 Cylindrical Cells

Cylindrical cells are typically referred to by their dimensions; for example, an 18650 cell refers to a cell that has a diameter of 18 mm and a length of 65 mm. Usually the metal can is the nega-

**Figure 1.4** Top section of a commercially available prismatic cell.

tive terminal of the cell with the positive terminal and vent at one end. Because the case of the cell is energized, the cells are partially covered with shrink-wrap for electrical isolation. Cylindrical cells use a single jelly roll of electrodes and separator and utilize one or more disconnect devices like a current interrupt device (CID) or a positive temperature coefficient (PTC) device for safety (the PTC is not commonly used in high-power cells). The can seal is typically formed through a crimp [3, 11].

Figure 1.5 shows a commercially available 18650 cell (a) and its top cap (b). Cylindrical Li-ion cells come in sizes up to 200 Ah. The end cap enclosure in an 18650 cell contains a gasket and a burst disk in the end cap assembly provides a venting mechanism in the cell. Although 18650 cylindrical cells are the most commonly used size cells, cylindrical cells are available in various sizes and are often packaged in the form of a battery pack for larger devices. For example, the 21700 type cylindrical cells are used in electric bikes. The 75400 type cells are often used in electronic cigarettes. The packaging density when grouping the larger format cylindrical cells in a pack is low due to their cylindrical shape [12].

### 1.4.3 Pouch Cells

Pouch cells are enclosed in a flexible package made from a heat-sealable aluminum plastic film. One construction of pouch cells consists of a jelly roll with metallic tabs attached to the positive and negative electrodes and protruding from the pouch. Due to its flat configuration and flexible packaging, pouch cells have a higher energy density and can be produced in different sizes based on the application. This has resulted in pouch cells being increasingly used in portable consumer electronic devices as it allows the de-

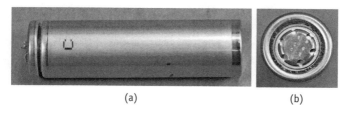

**Figure 1.5** (a) Profile view and (b) cap of a commercially available cylindrical cell.

signer to tailor the cell to the application rather than needing to tailor the application to the cell. Figure 1.6 shows a commercially available pouch cell with the enclosure cut off to reveal the electrodes. The electrodes and separator may be folded, stacked, or wound in pouch and prismatic cells. Pouch cells generally vent at the sealed edge of the pouch [11].

## 1.5 Battery Management System for Li-Ion Cells

Li-ion cells need to be operated within a well-defined set of conditions to prevent an elevated risk of cell failure during operation. For this reason, Li-ion batteries include a battery management system (BMS) that prevents the cell from operating outside its rated specifications. The battery management system, sometimes called the battery management unit (BMU), is a significant component in a battery pack, which not only acts as the interface between the Li-ion cells and the device/application but also more importantly provides protection to the cells and manages its operation throughout the life cycle of the cells in the device.[2] BMSs serve a variety of functions in a Li-ion battery system and the functionality of the BMS depends on factors such as the application, the capacity of the battery, and the exact Li-ion cell chemistry used. Chapter 5 discusses the BMS systems used in Li-ion battery systems.

## 1.6 Charger Circuits for Li-Ion Cells

Li-ion cells require a very specific algorithm for charging. Li-ion battery systems are often designed with sophisticated charging circuits that communicate with the BMS in the battery pack and can

---

2. In the authors' experience, the terms BMS, BMU, and PCM are commonly used interchangeably although the authors are not aware of a formal definition differentiating these terms. The term PCMs is typically used in the context of single-cell battery packs. PCMs provide overcharge, overdischarge, and overcurrent protection. The term BMU is more commonly used for multicell batteries such as those used in laptops or tablets. The BMU includes the overcharge, overdischarge, and overcurrent protection that is provided by the PCM but also includes additional features such as state of charge determination, the ability to communicate with a charger in the host device (typically via the I2C bus etc). BMS while often used in the context of multi-cell batteries is more commonly in the authors' experience used in the context of much larger batteries (e.g., automotive battery systems) and includes additional features such as state of health determination and so on.

**Figure 1.6** Commercially available pouch cell.

be used to identify the battery pack and perform a number of other functions. Chapter 4 discusses Li-ion charging circuits in detail.

## 1.7 End of Life

End of life (EOL) in the context of Li-ion cells refers to a condition where the cell can no longer serve its function in an application. For Li-ion cells it is often useful to think of its useful lifetime in terms of charge/discharge cycles with the expectation that a Li-ion cell will function for a specific number of charge and discharge cycles before reaching its end of life. Commonly, end of life is defined using some measure of faded performance relative to beginning of life. For a given application, the Li-ion battery should be sized such that the power/energy requirements of the application are met throughout the expected life expectancy of the product. EOL criteria differ for energy versus power applications and for individual cells versus complete battery pack systems. EOL for energy applications is commonly taken as when the battery's usable energy drops to 70% to 80% of the original energy. The rate of capacity deterioration typically increases once the capacity has dropped to approximately 70%. Usable energy is primarily influenced by capacity loss but may also be influenced by power fade (power fade occurs when the battery or cells are not able to deliver the current at the specified voltage). Determining whether a Li-ion cell or battery is suitable for the application requires identifying the EOL criteria for the cell/battery and predicting the capacity degradation of the cell/battery due to both cycle and calendar aging.

It is important to remember that Li-ion cells will degrade even when the cells are not being charged or discharged and just sitting on a shelf. This is known as calendar aging. As an example, the capacity loss at 60°C for a particular type of 18650 cells stored at different states of charge is as shown in Figure 1.7 [13]. The data indicates that a Li-ion cell may lose over 20% of its rated capacity if stored at 60°C in a fully charged condition even when the Li-ion cell is not used. The ambient temperature at which a cell operates and/or is stored can have a large impact on the cell's capacity degradation. A high ambient temperature during storage coupled with a high state of charge (SOC) accelerates the rate of capacity degradation in a cell.

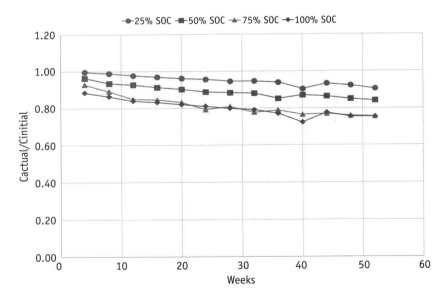

**Figure 1.7** Normalized capacity over 52 weeks at 60°C for an 18650 cell at different states of charge.

## 1.8 Shelf Life

Shelf life[3] expresses the theoretical lifetime of a cell when at rest at a given temperature and state of charge (SOC) and generally expresses the available capacity of a cell when it is returned to service after being stored. Multiple degradation mechanisms are typically involved in the loss of a Li-ion cell's capacity when at rest. For Li-ion chemistries with graphitic negative electrodes, the growth of the SEI layer is usually a dominant degradation mechanism. SEI layer growth increases the cell's impedance and reduces its capacity as it consumes cycleable lithium from the system [14]. The two main factors that affect shelf life of Li-ion cells are

- SOC at which the cells are stored;
- Storage temperature.

The shelf life of a particular Li-ion cell will depend on the actual design and electrochemistry of the cell. Several models have been developed to characterize the shelf life of Li-ion cells at different storage temperatures and states of charge. As an example, Table 1.3 shows the approximate capacity degradation for a lithium nickel cobalt aluminum oxide (NCA) type Li-ion cell at different storage temperatures and SOCs [15]. Each cell type will have a different shelf life under different operating conditions. As such, it is important to test the cell being considered for an application to determine its shelf life and suitability for an application.

**Table 1.3**

Approximate Capacity Degradation for a NCA Type Li-Ion Cell after 5 Years as a Function of State of Charge and Storage Temperature

| Storage Temperature (°C) | 50% SOC | 100% SOC |
|---|---|---|
| 20 | 5% | 13% |
| 40 | 10% | 25% |
| 60 | 20% | >35% |

---

3. Also commonly called calendar life.

The data in Table 1.3 indicates that the capacity degradation for the tested NCA type Li-ion cell after 5 years increases by

- Approximately 250% as the SOC during storage increases from 50% to 100%;
- Approximately doubles for every 20°C increase in storage temperature.

Reference [15] also indicates that the rate of capacity degradation during storage for the tested cell type slows down with time:

- For a SOC of 50% and a storage temperature of 40°C, the capacity degradation will be approximately 4% after 1 year and 10% after 5 years;
- For a SOC of 100% and a storage temperature of 40°C, the capacity degradation will be approximately 10% after 1 year and 25% after 5 years.

### 1.8.1 Life Test and Verification Process

Numerous models and prediction techniques have been proposed in the literature to predict the life of a Li-ion battery in a specific application. Aging tests are typically performed to characterize the capacity degradation of Li-ion cells in various applications. The aging tests are used to develop a model of a cell's capacity degradation in the field that is then used for predicting a cell's capacity degradation in an application. Typically, the testing process involves the following:

- Quantification of the expected duty cycle and environment.
- Definition of the EOL criteria.
- Creation of a matrix of aging test profiles that includes temperatures, charge/discharge currents, states of charge, and depth of discharge.
- Definition of the intervals for reference performance tests. The aim of these tests is to map changes in cell capacity during the aging test at periodic intervals.

- Aging experiments, which are typically performed for a few months (sometimes longer). The duration is dependent on the required predictive confidence level.
- Fitting a life model to the acquired data during the aging test. Typically, regression is used to determine semiempirical models for life prediction in the application.

The Arrhenius equation is commonly used for characterizing the temperature dependency of capacity degradation during storage. The aging factor is given by the following equation:

$$\text{Aging factor } \alpha \propto \exp\left(-\frac{E_A}{RT}\right)$$

where $E_A$ is the activation energy, $R$ is the gas constant, and $T$ is the temperature in degrees K.

The aging factor typically varies by cell type and can be determined experimentally by performing accelerated aging tests at different temperatures to characterize the capacity degradation as a function of time and temperature. The aging factor is determined for a particular SOC. Once determined, the aging factor can be used to predict the shelf life of a Li-ion cell at a particular SOC for different temperatures.

## 1.9 Cycle Life

Cycle life is a term associated with the aging of a Li-ion cell as it is charged and discharged. The cycle life of a Li-ion cell will be dependent on the specific cycling profile used in an application and the severity of the charge and discharge cycles. As an example, Figure 1.8 shows the capacity degradation as a function of charge/discharge cycles for a Li-ion cell with a lithium cobalt dioxide positive electrode and a graphite negative electrode. The figure indicates that the cell loses approximately 20% of its rated capacity after 300 charge/discharge cycles under the tested conditions at an ambient temperature of 25°C.

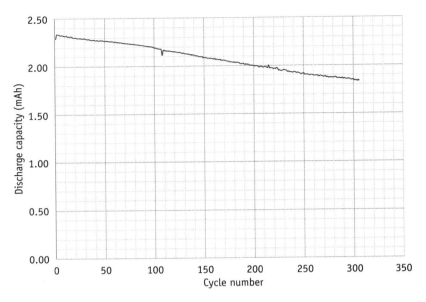

**Figure 1.8** Capacity as a function of charge/discharge cycles of a commercially available Li-ion cell.

# References

[1] Johnson Matthey Battery Systems, *Our Guide to Batteries*, 2012, http://www.jmbatterysystems.com/JMBS/media/JMBS/Documents/JMBS-Guide-to-Batteries.pdf.

[2] Naoki, N., et al., "Li-ion Battery Materials: Present and Future," *Materials Today*, Vol. 18, No. 5, June 2015.

[3] Reddy, T. B. (ed.), *Linden's Handbook of Batteries*, Fourth Edition, New York: McGraw Hill, 2011.

[4] Mikolajczak, C., M. Kahn, K. White, and R. T. Long, *Lithium-ion Batteries Hazard and Use Assessment*, New York: Springer Science & Business Media, 2012.

[5] van Schalkwijk, W., and B. Scrosati, *Advances in Lithium-Ion Batteries*, New York: Springer Science & Business Media, 2002.

[6] Arora, A., N. K. Medora, T. Livernois, and J. Swart, Safety of Lithium-Ion Batteries for Hybrid Electric Vehicles. In G. Pistoia (ed.), *Electric and Hybrid Vehicles, Power Sources, Models, Sustainability, Infrastructure and the Market*, Amsterdam: Elsevier, 2010, Chapter 18, pp. 463–491.

[7] Nitta, N., et al., "Li-Ion Battery Materials: Present and Future," *Materials Today*, Vol. 18, No. 5, 2015, pp. 252–264.

[8] Santhanagopalan, S., et al., *Design and Analysis of Large Lithium-Ion Battery Systems,* Norwood, MA: Artech, 2015, p. 12.

[9] IEEE Std. 1725.

[10] Santhanagopalan S., et al., *Design and Analysis of Large Lithium-Ion Battery Systems,* 2015 Artech, p. 91.

[11] Weicker, P., *A System's Approach to Lithium-Ion Battery Management,* Norwood, MA: Artech, 2014.

[12] Maiser, E., "Battery Packaging—Technology Review," *AIP Conference Proceedings,* Vol. 1597, No. 204, 2014, doi: 10.1063/1.4878489.

[13] Lele, S., et al., Predicting the life of LI-ion batteries using the Arrhenius model, Battcon 2018 International Battery Conference, Nashville, TN: April 22–25, 2018.

[14] Smith, K., Y. Shi, and S. Santhanagopalan, "Degradation Mechanisms and Lifetime Prediction for Lithium-Ion Batteries: A Control Perspective," 2015 American Control Conference Chicago, Illinois, July 1–3, 2015, p. 1.

[15] Santhanagopalan, S., et al., *Design and Analysis of Large Lithium-ion Battery Systems,* Norwood, MA: Artech House, 2015, p. 84.

[16] *Lithium Ion Rechargeable Batteries Technical Handbook,* Sony (undated).

[17] https://na.industrial.panasonic.com/sites/default/pidsa/files/panasonic_liion_battery_product_information_sheet.pdf.

[18] Wang, Q., B. Jiang, B. Li, and Y. Yan, "A Critical Review of Thermal Management Models and Solutions of Lithium-Ion Batteries for the Development of Pure Electric Vehicles," *Renewable and Sustainable Energy Reviews,* Vol. 64, 2016, pp. 106–128.

# 2

# Overview of Li-Ion Battery Systems

The many advantages of Li-ion batteries over other competing battery technologies, such as NiMH or lead acid, are well documented in the literature and were briefly discussed in Chapter 1. Li-ion battery systems have already become the leading choice for energy storage technology in a variety of applications. Like all typical battery systems, Li-ion battery systems also require an effective management of the stored energy to ensure both safety and reliability. Li-ion cells have a restricted range of operating current, voltage, and temperature. Operation outside the specified range of conditions can be detrimental to safety, reliability, and service life. Applications utilizing Li-ion batteries incorporate sophisticated charging and protection circuits to ensure that the cells are always operated within their rated specifications. Various Li-ion battery system designs exist that ensure a safe and reliable operation of the Li-ion cells in an application. The design and overall architecture of the Li-ion battery system depends on the application and its requirements. The design and architecture of Li-ion battery systems in grid storage applications and electric vehicles are very different from the design and architecture of the Li-ion battery systems in portable consumer electronic devices.

The term energy storage system (ESS) is typically used for systems that store energy using a variety of means including thermal, electromechanical, pumped hydro, or electrochemical. All ESSs capture energy and store it for use at a later time. Battery energy storage systems (often abbreviated as BESS) are a subset of ESSs with Li-ion energy storage systems being a type of a BESS [1]. In the literature, the term "Li-ion energy storage system" is often used in the context of large Li-ion battery systems, such as those used in grid storage applications. Large Li-ion battery storage systems can have a capacity range of up to tens of megawatts for short-and medium-duration applications with a life of 15 years and efficiency as high as 95%. However, this term is also applicable to all types of systems utilizing Li-ion batteries to store energy.

IEEE Std. 1625 describes a Li-ion battery system[1] as a "combined cell, battery pack, host device, power supply or adapter, end user and environment." [2] All Li-ion battery system designs (whether for portable consumer electronic devices or large scale grid-based energy storage systems) typically incorporate all these elements. However, the requirements and overall design of the Li-ion battery system and its various subsystems are specific to the application.

## 2.1 Single-Cell Li-Ion Battery Systems

In this book, a single-cell Li-ion battery system is considered as a system with one or more Li-ion cells connected in parallel. These systems typically power portable devices, such as cell phones, baby monitors, small toys, and wearable devices. Figure 2.1 shows the various subsystems that are part of a typical single-cell Li-ion battery system. The subsystems perform the following functions:

- *Alternating current (AC) adapter (or power supply):* The AC adapter functions as a power supply for the system. In a

---

1. The term Li-ion energy storage system and Li-ion battery system are used interchangeably in this book.

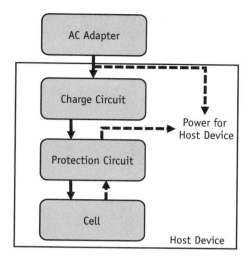

**Figure 2.1** Single-cell Li-ion battery system. The solid arrows show the direction of the charge current and the dotted arrows show the current that is used to power the system.

typical application, the AC adapter converts the AC voltage from a wall outlet (120 Vac in the United States) into a direct current (DC) voltage that is used to power both the host device and also charge the Li-ion battery. A Universal Serial Bus (USB) type adapter is a common adapter type used in single-cell Li-ion battery systems. AC adapters are discussed in detail in Chapter 3.

- *Host device*: The host device is powered by the battery when the AC adapter is no longer connected to the device. The host device also typically incorporates a charge circuit to provide a charge current to the battery when it is connected to an external power source.
- *Charge circuit*: The charge circuit is typically located in the device that is powered by the Li-ion battery. The charge circuit conditions the DC voltage from the AC adapter to an appropriate voltage and current to charge the Li-ion battery. In addition to providing the appropriate charge voltage and current to the Li-ion battery, the charge circuit also typically serves a number of other functions, such as
  - Monitoring the battery's state of charge (SOC);

- Monitoring the battery pack's ambient temperature to ensure that the battery is only charged when the battery temperature is within its rated temperature range;
- Monitoring the battery pack's voltage to ensure that a preconditioning current (precharge) is provided to an overdischarged cell;
- Communicating with the battery to determine the voltage/current to use for charging the battery (this is not very common in single-cell applications);
- Determining the identity of the battery to ensure that only the correct battery is being charged;
- Monitoring the charge current and charge time to ensure that current is not supplied to a faulty battery pack.

Charge circuits are discussed in detail in Chapter 4.

- *Protection circuit*: The protection circuit (also commonly called the protection circuit module or PCM) together with the cell forms the Li-ion battery. For single-cell Li-ion batteries, the PCM is typically attached directly to the cell terminals. The role of the PCM is to prevent the cell from operating outside its rated specifications. Single-cell Li-ion battery system PCMs typically provide overcharge, overcurrent, and overdischarge protection. Should the charger or the load's low voltage cut-out fail, the PCM offers a redundant protection to prevent a single point of failure from resulting in the operation of the cell outside its specifications. Battery protection circuits are discussed in detail in Chapter 5.
- *Cell:* The cell stores the electrical energy that is used to power the system when the AC adapter is disconnected from either the wall outlet or from the system. Single-cell Li-ion battery systems also often incorporate passive protection devices, such as positive temperature coefficients (PTCs), thermal fuses, bimetallic, and switches. These passive protection devices provide backup protection to the cell in the event of a failure of the PCM.

Applications such as portable USB power banks and wearable devices utilizing a single Li-ion cell or multiple Li-ion cells connected in parallel sometimes may not have a PCM connected directly to the cell terminals. The PCM in these devices can be located on the device's circuit board. The cell terminals in these devices are connected directly to the circuit board. Some e-cigarette designs use a single cylindrical (18650) Li-ion cell without a battery protection circuit attached either to the cell terminals or incorporated in the e-cigarette itself. The lack of a protection circuit for the Li-ion cells makes the cells used in these e-cigarettes susceptible to failure due to a fault on the e-cigarette's circuit.

## 2.2 Multicell Li-Ion Battery Systems

Multicell Li-ion battery systems may include several cells connected both in series and parallel. Figure 2.2 shows the various subsystems that form part of a typical multicell Li-ion battery system in a portable consumer electronics application [3]. The various subsystems of the Li-ion battery system typically include the following:

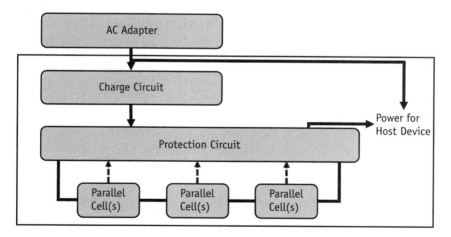

**Figure 2.2** Multicell Li-ion battery system in consumer electronics application. The solid lines depict current flow and the dotted lines depict voltage sense wires that communicate the voltage of each parallel cell combination to the protection circuit.

- *AC adapter (or power supply):* Similar to single-cell Li-ion battery systems, AC adapters in multicell Li-ion battery systems function as a power supply for the system and convert the input AC voltage into a DC voltage that is used to both power the host device and charge the battery. AC adapters are discussed in detail in Chapter 3.
- *Host device:* The host device is powered by the battery and also contains a smart charger circuit that provides a charge current for the battery when it is connected to an external power source, such as an AC adapter.
- *(Smart) charger:* The charger that is typically incorporated in the host device is designed to provide an appropriate charge voltage and current to the battery. Typically, chargers in small multicell Li-ion battery systems also perform the following functions:
    - Communicate with the battery to determine the battery's identification, state of charge, and temperature;
    - Prevent the battery from being charged outside a predetermined temperature range;
    - Monitor the charge current to the battery and detect a fault condition in the charge path (e.g., overcurrent, short-circuit).
- *Battery pack:* Multicell Li-ion battery packs[2] contain a battery protection circuit (BMU or BMS) that provides overcharge, overdischarge, and overcurrent protection to the cells in the battery pack. For multicell battery packs with cylindrical cells, the cells are often connected to each other through tabs. In addition, voltage sense wires often connect the cells to the battery pack's BMU. The BMU in multicell battery packs is typically designed with multiple levels of redundant protection. The BMU monitors the voltage of each individual parallel cell block in the battery pack and provides charge/discharge voltage, current control, and temperature control. Typically, BMUs in small multicell battery packs for portable consumer electronic devices do not monitor the temperature of each individual cell in the battery pack. The temperature

---

2. The terms battery and battery pack are used interchangeably in this book.

sensors are installed in selected locations in the battery pack to monitor the battery pack's temperature. It is not uncommon for BMUs in multicell battery packs to also incorporate cell balancing circuits. This is discussed further in Chapter 5.

### 2.2.1 Other Configurations

Unlike consumer electronic devices such as laptops and tablets, where the charge circuit is located on a circuit board in the device, applications such as electric bicycles, electric scooters, and hover boards sometimes incorporate both the charger and protection circuit inside the battery pack (on the battery protection circuit). An external power supply (AC adapter) providing a DC output voltage connects directly to the battery pack in these applications.

## 2.3 Automotive Battery Systems

There are many applications of Li-ion batteries in an automobile. These applications can range from batteries that are used to start the engine, batteries that are part of the hybrid electric vehicle (HEV), to batteries that provide the sole propulsion energy in electric vehicles (EVs) [4]. The number of cells in an HEV or EV application typically falls in the hundreds. Batteries in these applications are designed to provide a relatively high voltage and current. This requires that the BMS in these systems is comparatively more sophisticated than those in portable consumer electronic systems. A BMS structure with vehicle energy management functions is as shown in Figure 2.3 [5]. The figure demonstrates the various functions that may be performed by the battery management system in an EV battery pack. While a full discussion of these functions is outside the scope of this chapter, the figure demonstrates the sophisticated nature of the battery management system in larger battery packs.

A BMS in an automotive application typically provides the following functions:

- Ability to determine the battery's state of charge and state of health in real time (The concept of a battery's state of health is

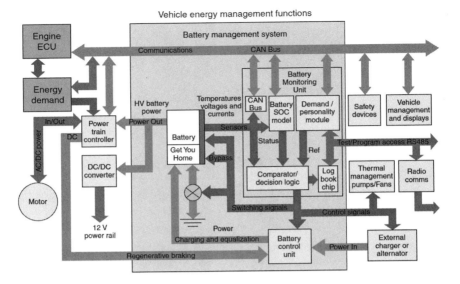

**Figure 2.3** Example protection circuit for a HEV battery [5].

one that attempts to classify battery degradation using a simple metric which indicates how far a battery has progressed towards its end of life [6];)

- Ability to control the charging and discharging of the cells to prevent the cells from operating outside their rated temperatures;
- Ability to balance the cells;
- Thermal management;
- Communication of battery status to a user interface.

As is common in BMUs used in portable consumer electronic products, the BMS in an EV or HEV application includes multiple levels of redundant protection for current, voltage, temperature, power, and other parameters.

## 2.4 Battery Systems for Large-Scale Energy Storage Systems

Li-ion battery systems are also extensively being used in the power grid industry not only to serve as a backup power source during outages, but to ensure stability of the grid's power (especially for renewable sources) by [7]:

- Using fluctuating power from renewable sources to charge the battery systems;
- Peak-load shifting at different times during the day. Consequently the high power demand of load peaks is deliberately shifted to times when there is a lower demand on the grid.

Large ESS in the multiple kilowatt range can also be connected to a multitude of solar panel arrays with the battery being charged using solar power. These inverters use state-of-the-art technologies with wide input voltages and high efficiencies exceeding 95%. Battery systems used for large-scale energy storage have major differences when compared to the single-cell or multicell Li-ion battery systems discussed in Sections 2.2 and 2.3. These systems typically utilize a larger number of cells (often numbering in the hundreds or even in the thousands) and can operate at significantly higher voltages even when compared to automotive battery systems. Compared to battery systems in portable consumer electronic devices where the total energy capacity of a battery may be in the 50- to 100-Wh range, the typical energy capacities for battery systems in grid applications are in the kilowatt-hours or even the megawatt-hours range with operating voltages commonly exceeding 1000 Vdc. Operating currents in these systems can be in the hundreds of amperes range (or even higher). The significantly higher operating voltages and currents introduce additional system requirements that are not commonly applicable to smaller battery systems in consumer electronic devices or even electric vehicles. The interaction between the battery system and the loads on the grid is more complex than battery systems in either consumer electronics devices or electric vehicles.

The design of the battery system and its BMS in larger-format battery systems depends on the level of modularity desired and can range from a simple monolithic system where all functionality is placed in a single module, to a distributed system with a high degree of modularity. A common design in large-format battery systems consists of a single-string architecture where the number of Li-ion cells needed to achieve a certain amp-hour capacity are placed in parallel to create a cell-group, essentially forming a large cell with a capacity equal to the capacity of the individual cell multiplied by the number of cells in parallel. The required number of series elements is then installed in series to achieve the required system voltage. A common battery system architecture includes a single central master control module that is responsible for the computational requirements, and a number of slave modules that connect directly to the cells to measure the cell voltage and temperature and communicate this information to the central control module [6].

## References

[1] *Lithium-Ion Battery Energy Storage Systems*, AIG Energy Industry Group.

[2] IEEE 1625-2008, IEEE Standard for Rechargeable Batteries for Multi-Cell Mobile Computing Devices, IEEE Power and Energy Society.

[3] Arora, A., et al., "Lithium-ion Batteries for Hybrid Electric Vehicles: A Safety Perspective." *5th International Advanced Automotive Battery (and Ultracapacitor) Conference*, Hawaii, 2005.

[4] Santhanagopalan, S., et al., *Design and Analysis of Large Lithium-ion Battery Systems*, Norwood, MA: Artech, 2015, p. 142.

[5] Arora, A., N. K. Medora, T. Livernois, and J. Swart, "Safety of Lithium-Ion Batteries for Hybrid Electric Vehicles." In G. Pistoia (ed.), *Electric and Hybrid Vehicles, Power Sources, Models, Sustainability, Infrastructure and the Market*, Amsterdam: Elsevier, 2010, Chapter 18, pp. 463–491.

[6] Weicker, P., *A Systems Approach to Lithium-Ion Battery Management*, Norwood, MA: Artech, 2014, p. 66.

[7] Evans, A., et al., "Assessment of Utility Energy Storage Options for Increased Renewable Energy Penetration," *Renewable and Sustainable Energy Reviews*, No. 16, 2012, pp. 4141–4147.

# 3

# Power Supplies for Portable Consumer Electronic Products

As discussed in Chapter 2, Li-ion battery systems, whether they are single-cell battery systems used in applications, such as cell phones and power banks, or large-scale battery systems, such as those used in grid storage applications, require a power source to charge the cells. For portable consumer electronic devices, this power source can be an AC power source, such as a wall outlet in the house, or a DC power source, such as a USB port in a vehicle. In larger battery systems, such as those used for energy storage in residential applications, the power source may be an inverter or a solar charge controller. The role of the power source (e.g., an AC adapter in a single-cell battery system or an AC-DC converter in a large battery system) is to convert the input voltage (whether AC or DC) into a DC voltage that can be used by the charge circuit in the battery system to charge the Li-ion cells. In portable consumer electronic devices, the power source when connected to the device also provides power to the device for operation. This chapter will focus on the power supply (AC adapter) used by portable consumer electronic devices. The chapter will also introduce DC-DC power supplies often used by portable consumer electronic devices. These power supplies typically step down a DC voltage to a lower voltage which is then supplied to the battery charge circuit in the device.

## 3.1 Components of a Typical AC-DC Converter (AC Adapter) Circuit

Most portable consumer electronic devices are sold with an AC adapter that converts the AC voltage from a wall outlet (120 Vac in the United States) to a DC voltage (e.g., 5 Vdc for USB adapters) to power the device and charge the battery in the device. AC adapters shipped with portable consumer electronic devices use switch-mode power supplies (SMPS). SMPSs come in many different varieties, with the flyback converter being a popular circuit used for AC adapter designs in portable consumer electronic devices. The common AC-DC SMPS types include [1]

- *Flyback converters*: commonly used in low-power applications (typically less than 100W) due to their simplicity and relatively lower cost;
- *Forward converters*: commonly used for intermediate power levels due to their better utilization of the transformer and lower active device peak current;
- *Full-bridge converters*: mainly used in high-power applications (typically greater than 1000W) due to their overall complexity and cost.

Figure 3.1 shows an example of a typical high-level design of the AC adapters shipped with consumer electronic devices. The main components of the system shown in Figure 3.1 are

- *Input overcurrent protection:* Fuses and/or PTCs are commonly used for overcurrent protection at the adapter input. The

## 3.1 Components of a Typical AC-DC Converter (AC Adapter) Circuit

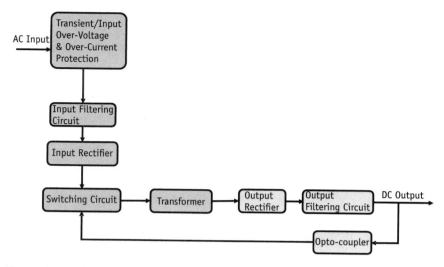

**Figure 3.1** High-level block diagram of an AC adapter design typically used with consumer electronic products. The dark gray boxes represent the high AC voltage sections of the adapter while the light grey boxes represent the low DC voltage sections of the adapter. The transformer and the optocoupler provide galvanic isolation between the primary high voltage section of the adapter and the secondary low voltage section.

role of the overcurrent protection element at the adapter input is to limit current flow to the adapter circuit in the event of a failure (e.g., a short-circuit failure of a component on the adapter circuit).

- *Transient/input overvoltage protection:* Transient voltage surge suppressors (TVSS) or metal oxide varistors (MOVs) are commonly used for providing transient overvoltage protection to the adapter circuit. The role of the TVSS/MOV is to clamp the magnitude of the transient by absorbing the transient energy on the AC input and preventing damage to components downstream in the adapter circuit. As opposed to a TVSS, MOVs can absorb higher transient energies and are inherently bidirectional in nature. A varistor (variable resistor) is an electronic component whose electrical resistance varies with the applied voltage. MOVs are commonly used for input surge protection in AC adapters. At rated voltages, MOVs have a high resistance, but at higher voltages the resistance drops significantly. MOVs are commonly made of a ceramic of zinc oxide grains in a matrix of other oxides. The grains form diodes with the surrounding matrix, creating a complex array of parallel and antiparallel diodes. At low voltages each diode has a small voltage across it resulting in minimal current flow. At high voltages, the individual diodes begin to conduct and the resistance of the MOV drops significantly. Factors such as grain size, the nature of the matrix material between the grains, the thickness of the ceramic, and the attachment of leads to the ceramic determine the overall properties of the MOV [2]. Figure 3.2 shows one example of a MOV's construction.

MOVs can be a fire hazard if they are subjected to a continuous abnormal voltage condition rather than short duration transients. If a MOV is subjected to a sustained abnormal overvoltage, the MOV may go into thermal runaway resulting in overheating, smoke, and potentially a fire [3]. Furthermore, MOVs degrade over time with the number of applied transients due to the fusing of grains resulting in a decrease in the maximum continuous voltage rating. This condition eventually results in electrical conduction within the MOV at the rated operating voltage, ultimately leading to a failure of the MOV. For this reason, MOVs require some level of protection against this failure mode. This protection is typically provided by a thermal fuse or thermal cutoff device that is installed in thermal contact with the MOV. This thermal

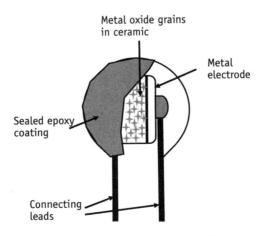

**Figure 3.2** Construction of a MOV.

fuse detects the MOV's temperature and removes power from the MOV once it starts to overheat. However, because the MOV is a shunt element, its failure as an open circuit results in a loss of surge protection for the adapter circuit. For this reason the thermal fuse is often put in series with the input power to disable the AC adapter circuit altogether in the event of a failure of the MOV.

- *Input filtering circuit:* The input filtering circuit may include a combination of capacitors and inductors. These components reduce the electromagnetic interference (EMI) between the AC adapter and the AC input power line, and also limits the effects of EMI on the AC adapter from the input power source that may interfere with the operation of the adapter.
- *Input rectifier:* The rectifier converts the AC input voltage into a DC voltage.
- *Power factor correction:* In switch mode power supplies, the power supply circuit may cause the input line current to be nonsinusoidal. Even if it is sinusoidal, it may be out of phase with the sinusoidal input voltage or have harmonics of the line voltage. This can result in a lowered power factor and a consequent waste of power (i.e., a less efficient power supply). For this reason, switch-mode power supplies can sometimes include a power factor correction circuit (this is not shown in Figure 3.1). The role of this circuit is to maximize the power factor by forcing the input line current to be sinusoidal and in phase with the input line voltage, resulting in fewer and lower line harmonics [1].
- *Transformer*: The transformer steps down the high voltage on the primary side of the adapter circuit to a lower voltage on the secondary side of the adapter. The transformer also provides isolation between the high voltage input section (primary side) of the adapter and the low voltage output section (secondary side).
- *Switching circuit*: At a high level, the switching circuit switches the current flow through the transformer on the primary side to enable the transformer to step down the voltage to a lower level on the secondary side. The switching circuit mod-

ifies its switching duty cycle based on the feedback it receives regarding the magnitude of the adapter output voltage from the optocoupler. The switching circuit may also be connected to a thermistor that monitors the internal temperature of the adapter and can turn off the adapter if the temperature exceeds a preset threshold.

- *Output rectifier/output filtering circuit:* The output rectifier together with the output filtering circuit converts the switched DC voltage on the secondary side of the transformer into the DC output voltage to power the device.
- *Optocoupler:* The optocoupler monitors the adapter's output voltage and provides feedback information to the switching circuit on the primary side of the adapter. The switching circuit uses this information to modify its switching characteristics and ensure that the adapter's output voltage is at the designed magnitude. The optocoupler also provides isolation between the high and low voltage sections of the adapter.
- *Secondary output overvoltage protection:* AC adapters also often incorporate an independent secondary output overvoltage protection circuit (not shown in Figure 3.1). The role of this circuit is to provide independent secondary protection in the event of a failure of the switching circuit or the transformer. A zener diode is often used to provide this protection in some adapters used in portable consumer electronic devices.

## 3.2 AC Adapter Requirements[1]

The electrical circuit in an AC adapter includes the relatively high AC input voltage and the lower DC output voltage. The presence of the high AC input voltage makes the adapter susceptible to issues, such as propagating circuit board failures, overtemperature conditions, overcurrent conditions, and electric shock hazards.

---

1. This section will discuss the requirements for power supplies (AC adapters) used in portable consumer electronic devices.

For this reason, AC adapters must be designed and constructed to minimize the potential of these hazards.

### 3.2.1 Propagating Circuit Board Failures

The failure of a printed circuit board (PCB) may be due to [4]

- The design or manufacturing of the PCB itself;
- Conduction between components or traces on the PCB that were designed to be electrically isolated;
- A component on the circuit board that only affects the board;
- A component whose failure affects surrounding components;
- A component that results in the propagation of the failure.

A propagating PCB failure in an AC adapter can result in a safety concern. The initiating mechanism for a propagating PCB failure is the formation of a resistive path between two traces or layers at different potentials. This resistive path can form due to an external heat source, insulation breakdown, poor construction/soldering techniques, arcing, or contamination [5]. A propagating PCB failure typically starts slowly from an initial resistive path between two conductors at different electric potentials. The heat dissipated through the resistive path pyrolizes the insulating material. The pyrolized insulating material is electrically a poor conductor and adds to the heat generated by the original resistive fault. The charred resistive fault continues to grow and propagates toward the source of the power, feeding the fault. Typically the input power connection on an AC adapter PCB is the area that is most susceptible to a propagating PCB failure. Poor control during the assembly process (e.g., poor soldering, introduction of contaminants) in this area increases the risk of a failure. Figure 3.3 shows an example of a PCB in an AC adapter for a portable consumer electronics device that has a comparatively higher risk of a propagating PCB failure due to the inconsistent soldering of components. The area around the power input is especially susceptible to a failure due to the poor soldering and the presence of contaminants on the circuit board.

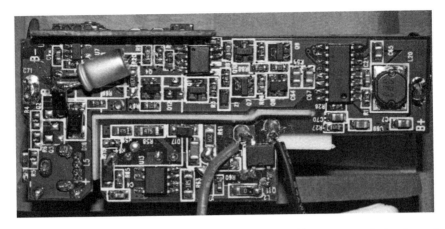

**Figure 3.3** The circuit board of an AC adapter for a portable consumer electronic product. This circuit board shows significant issues due to poor assembly/soldering. The poorly controlled assembly process results in inconsistently soldered components/wires, contaminants, and residue on the circuit board.

AC adapter circuits should be designed to reduce the risk of a fault on the circuit board since this can lead to a propagating PCB failure that results in a safety concern.

### 3.2.2 Burn Hazards

An overheating AC adapter can lead to a burn hazard to consumers. There are numerous instances of AC adapters that have been recalled because of the risk of burn hazards [6]. As an example, Figure 3.4 shows the temperature on the outside surface of an AC adapter under normal operating conditions. As can be seen in Figure 3.4, the adapter enclosure temperature can increase substantially when the airflow around the adapter is restricted (at approximately 4 hours and 45 minutes in the plot). Understanding the peak enclosure temperatures under worst case operating conditions and evaluating the risk of a burn hazard is important before an adapter design is finalized and mass production is started.

#### 3.2.2.1 Skin Temperatures

Determining whether an AC adapter can pose the risk of a burn hazard involves more than just characterizing the temperature of

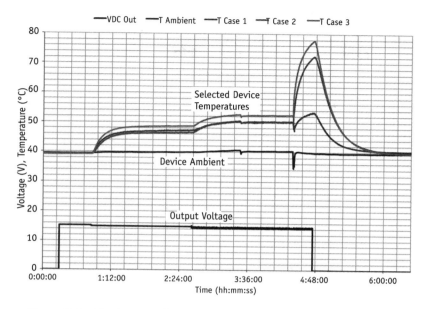

**Figure 3.4** AC adapter surface temperatures under a restricted air flow test.

the adapter's hot spot[2] under both normal operating conditions and fault conditions. When a hot object is brought in contact with a cold object, heat flows from the hot object to the cold object. The rate of the heat flow depends on a number of factors, including the difference in temperature between the two objects and the thermal resistance of each object. Theoretically, the heat flow will continue until the two objects are at thermal equilibrium (i.e., at the same temperature).

When a user touches a hot spot on the surface of an AC adapter, whether it is operating normally or has experienced a failure, the heat will flow from the adapter's enclosure to the skin, causing the skin temperature to rise. The skin temperature does not rise immediately to the temperature of the contacted adapter enclosure area. Rather, the rate of rise of skin temperature in this scenario depends on factors that include the temperature of the hot spot, the rate of heat flow from the hot spot to the skin, the heat stored in the hot spot, the duration of contact with the hot

---

2. A hot spot in this chapter refers to an area on an adapter's enclosure that exhibits temperatures that are comparatively higher than on other areas of the enclosure.

spot, the contact pressure, and the rate of heat carried away by the blood in the body. As an example, a metal at an elevated temperature feels hotter when touched than does a plastic at the same temperature. This is because metals have a comparatively lower thermal resistance than plastics, allowing a higher rate of heat transfer to the skin.

When a user touches a hot spot, the outer layer of the skin (epidermis) makes contact with the hot spot, which causes the skin temperature to rise. As the skin temperature increases, the skin suffers first-degree burns, followed by second-, and finally third-degree burns. Table 3.1 lists approximate skin temperatures for the different degrees of skin burn.[3]

#### 3.2.2.2 Allowable Surface Temperatures: Industry Standards

Several standards provide guidance on the maximum safe temperature for various materials. As an example, the European standard BS EN 13202:2000 provides tables relating surface temperature and contact time to burn threshold for a variety of materials (Table 3.2).

Another industry standard for information technology equipment, UL 60950-1, provides touch-temperature limits for parts in operator access areas (Table 3.3) [10].

Although the various industry standards provide useful guidelines for determining maximum acceptable surface temperatures, the diversity of materials and configurations in AC adapters may require additional investigation to determine the correct threshold

Table 3.1
The Various Degrees of Skin Burn

| Degree of Burn | Skin Temperature (°C) | Effect |
| --- | --- | --- |
| 1st | 44–55 | Damage to outer layer of skin. Burn heals on its own. |
| 2nd | 55–60 | First layer of skin burns through, and damage to the second layer of skin occurs. However, burn does not pass through to underlying tissues. |
| 3rd | >60 | Involves damage to all layers of the skin. |

From: [7, 8].

---

3. The numbers provided in this table are approximate values and do not account for the contact period.

**Table 3.2**
Burn Threshold Spreads for a Contact Period of 0.5 Seconds

| Material | Burn Threshold Spread for a Contact Period of 0.5 s (°C) |
|---|---|
| Bare (uncoated) metal | 67–73 |
| Ceramics, glass, and stone | 84–90 |
| Plastics | 91–99 |
| Wood | 128–155 |

From: [9].

**Table 3.3**
Maximum Temperature Specified in UL 60950-1, 2007

| Parts in Operator Access Areas | Maximum Temperature (°C) | | |
|---|---|---|---|
|  | Metal | Glass, Porcelain and Vitreous Material | Plastic and Rubber |
| Handles, knobs, grips, etc., held or touched for short periods only | 60 | 70 | 85 |
| Handles, knobs, grips, etc., continuously held in normal use | 55 | 65 | 75 |
| External surfaces of equipment that may be touched | 70 | 80 | 95 |
| Parts inside the equipment that may be touched | 70 | 80 | 95 |

for surface temperatures and characterize the potential for burn hazards.

#### 3.2.2.3 Experimentally Assessing Potential Burn Hazards

The American Society for Testing and Materials (ASTM) has published the *Standard Guide for Heated System Surface Conditions that Produce Contact Burn Injuries*, which provides a process for determining acceptable surface operating temperatures [11]. This standard includes a chart that relates contact skin temperature to the time needed to cause a burn. The curve labeled "Threshold B" in this chart represents the threshold for first degree burns. As an example, the chart indicates that a hot spot that causes the skin to attain a surface temperature of 60°C can produce first-degree burns after approximately 3 seconds.

Another ASTM standard, "Standard Practice for Determination of Skin Contact Temperature from Heated Surfaces Using a Mathematical Model and Thermesthesiometer," provides a procedure for evaluating the skin contact temperature for heated surfaces [11]. This standard details two procedures: a purely mathematical approximation technique that is typically used during the design phase of a product and the use of an instrument called a thermesthesiometer that "analogues the human sensory mechanism."

The mathematical model described in the standard "approximates the transient heat flow phenomena of the skin contacting a hot surface" and relies on a set of inputs to characterize the burn hazard. The model's output is only as good as the set of inputs, and the model requires a careful analysis of the system geometry, materials, operating temperatures, and air-flow measurements. The thermesthesiometer provides a comparatively simpler procedure for evaluating the potential of a burn hazard.

The two ASTM standards are typically used in conjunction with each other and allow for a modeling of the complex temperature-time relationship for burn injuries.

### 3.2.2.4 Thermesthesiometer

A thermesthesiometer is a device that measures heat flow and uses material with thermal characteristics similar to those of human skin. It is designed to provide an electrical analog (i.e., a simulation) of the finger's thermal response when placed against a heated surface [12]. A fine thermocouple wire is positioned approximately 0.01 cm below the material that simulates skin in the device, to sense temperature [11]. The thermesthesiometer also includes both an analog and a digital processing circuit to amplify the thermocouple signal, provide a temperature control circuit, process the thermocouple measurement, and display the temperature of the skin in degrees Celsius.

### 3.2.2.5 Case Study: Characterizing a Burn Hazard Using a Thermesthesiometer

The input supply voltage circuit in a consumer electronics device was failing in the field under a set of specific operating conditions. The failure of this circuit resulted in elevated temperatures on the surface of the device. Testing performed in the laboratory to re-create the failure indicated that the device's screen and area around the battery pack were the hottest user-accessible surfaces, and the temperature of both surfaces exceeded 120°C during the fault condition (Figure 3.5).

The thermesthesiometer was used to characterize the skin temperature in the event that a user contacted the screen and/or the area around the device's battery pack during a failure of the device. Measurements performed using a thermesthesiometer together with the temperature-time relationship curve from the ASTM standard, indicated that no irreversible epidermal injury would occur if the user released the overheated area around the battery pack within 8 seconds or the overheated screen within approximately 12 seconds (Figure 3.6). The data obtained from the experiment was then used to characterize the risk of the failure leading to a burn injury in the field.

Although the thermesthesiometer is a useful tool for characterizing burn hazards, it has some limitations. These include

- Its inability to deform about a device like the human skin;

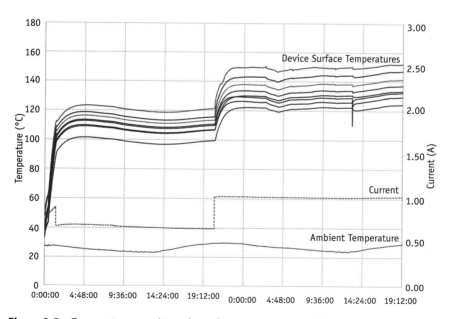

**Figure 3.5** Temperatures on the surface of a product under a fault condition of the input supply voltage circuit.

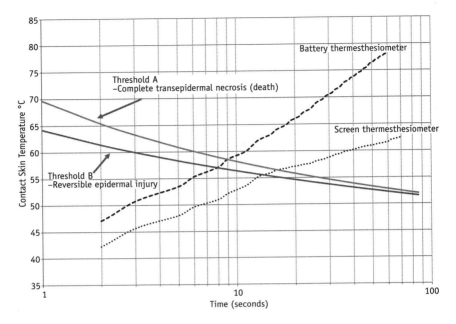

**Figure 3.6** Time to irreversible epidermal injury.

- Its inherent inaccuracy when used on surfaces that are uneven, rough, or too small for the probe to make full contact;
- The dependence of its accuracy on applying the right contact pressure to the surface of the heated contact area.

The above limitations must be kept in mind when reviewing the results obtained using the thermesthesiometer and performing a risk analysis.

### 3.2.3 Component Derating

Ensuring that components are adequately rated for an application is especially important in AC adapters as a component failure has the potential to lead to a propagating circuit board failure creating a safety hazard. When selecting components for use in AC adapters it is important to ensure that the components are rated to handle all foreseeable operating conditions (including abnormal conditions) in the field. For example

- At a high level, the function of a fuse is to provide fault protection by reliably melting under current overload conditions. Fuses have both a breaking capacity rating and a current rating. The breaking capacity rating of the fuse is the maximum current that the fuse can break at its rated voltage. The current rating of the fuse on the other hand is the value of current that the fuse can carry under a specific set of conditions [13]. When selecting a fuse for overcurrent protection, a determination should be made of all the different fault conditions that can exist to ensure that the fuse can provide protection under all fault conditions. As an example, Figure 3.7 shows a fuse that was not rated to handle all fault currents that were possible in a tested AC adapter. Although the fuse in the tested adapter could terminate the fault current in the event of a failure on the input filtering circuit, it was not adequately rated to terminate the fault current if a failure occurred on the switching circuit downstream from the filtering circuit. This fault resulted in a current which the fuse was unable to terminate before sustaining damage, resulting in a safety hazard.

- When selecting the temperature rating of components, care should be given not only to the expected temperatures under normal conditions, but also the temperatures that the component may see under all foreseeable operating conditions. As an example, Figure 3.8 shows the measured temperature of an electrolytic capacitor that is part of the input rectifier circuit in a tested adapter. In this adapter, the capacitor (C2 in

**Figure 3.7** An inappropriately rated fuse in a power supply.

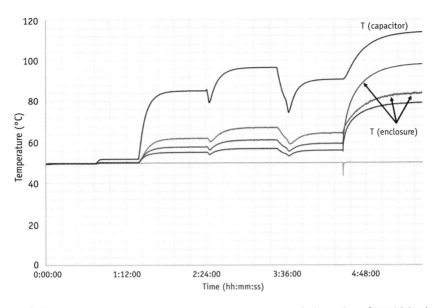

**Figure 3.8** The electrolytic capacitor C2 temperature exceeds its rating of 105°C in the tested power supply when a restricted air flow condition is simulated.

Figure 3.8) exceeds its rated temperature of 105°C when a restricted airflow condition is simulated. The operation of the capacitor above its rated temperature increases the probability of a premature capacitor failure, which in some instances can result in a safety hazard. Electrolytic capacitors commonly used in AC adapters for portable consumer electronic devices are typically rated for either 85°C or 105°C. The operating temperature of the capacitor impacts the capacitor's life in the application. As a general rule of thumb every 10°C reduction in the operating temperature of the electrolytic capacitor doubles the operating life of the capacitor in the application assuming all other operating parameters remain the same. The temperature of the adapter enclosure in this scenario also exceeded 90°C giving rise to the concern of a potential burn hazard.

- When selecting the current rating of components, care should be taken to ensure that the components will not fail and initiate a propagating circuit board failure in the event of a fault on the circuit board. As an example, Figure 3.9 shows

**Figure 3.9** Thermal damage to current sense resistors in a power supply application.

the thermal failure of current sense resistors used to monitor the current through the switching circuit in an AC adapter. Testing performed indicated that the current sense resistors failed before the fuse at the input could trip to terminate the fault current due to a short-circuit failure of the switching transistor.

### 3.2.4 Power Supply Efficiency

The regulations surrounding power supply efficiency and no-load power draw by power supplies has evolved rapidly since the California Energy Commission (CEC) implemented the first mandatory standard in 2004. A new set of requirements which went into effect in February 2016 were published by the Unites States Department of Energy (DOE) in 2014.

Power supply manufacturers typically indicate compliance to the efficiency requirements by placing a roman numeral on the power supply label, as specified by the International Efficiency Marking Protocol for External Power Supplies Version 3.0. Table 3.4 summarizes the thresholds for no-load power requirements and average efficiency requirements [14]. The internationally approved test method for determining efficiency has been published by International Electrotechnical Commission (IEC) as AS/NZS 4665 Part 1 and 2. Efficiency is established by measuring input and output power at four defined points: 25%, 50%, 75%, and 100% of rated power output. Typically, data for all four points is reported separately and an arithmetic average is used to determine the average active efficiency.

### 3.2.5 Single Points of Failure

The AC adapter circuit should be designed such that the adapter output voltage under any single fault condition in the circuit does not lead to an overvoltage condition at the output. An overvoltage condition at the adapter's output can lead to a cascading failure in the battery system possibly resulting in an overcharge of the Li-ion cell(s). As discussed in Section 3.1, AC adapters often include a zener diode at the output, which provides independent output over-voltage protection and limits the adapter output voltage in the event of such a failure.

Another component sometimes used for output overvoltage protection is a PolyZen device.[4] In a PolyZen device a zener diode is thermally coupled to a resistively nonlinear polymer PTC layer. The integrated PTC layer is electrically in series with the input node and the diode clamped output node. The polymer PTC layer in the PolyZen device responds to either extended diode heating and/or overcurrent events by transitioning from a low to high resistance state. This limits the current and also generates a voltage

---

4. The term PolyZen is a trademark of Littlefuse. Other manufacturers use different names for similar devices. For example, Eaton makes a similar component, which it calls a 'PolySurg'. These components are also sometimes referred to as mixed technology transient voltage suppressor devices.

## Table 3.4
### No-load Power and Average Efficiency Requirements

| Level | No-Load Power Requirement | Average Efficiency Requirement |
|---|---|---|
| I | Used if you do not meet any of the criteria | |
| II | No criteria was ever established | |
| III | Power rating ≤10 watts: ≤ 0.5W of no load power<br>Power rating: 10-250 Watts: ≤ 0.75W of no load power | • Power Rating ≤ 1W: Efficiency requirements – Power Rating × 0.49<br>• Power Rating 1-49W: Efficiency requirements ≥[0.09 × Ln(Power Rating)]+0.49<br>• Power Rating 49-250W: Efficiency requirements ≥84% |
| IV | Power rating: 0-250 watts: ≤ 0.5W of no load power | • Power Rating ≤ 1W: Efficiency requirements – Power Rating × 0.50<br>• Power Rating 1-51W: Efficiency requirements ≥[0.09 × Ln(Power Rating)]+0.5<br>• Power Rating 51-250W: Efficiency requirements ≥85% |
| V | **Output Voltage > 6V**<br>Power rating 0-49 watts: ≤ 0.3W of no load power<br>Power rating: 50-250 watts: ≤ 0.5W of no load power | • Power Rating ≤ 1W: Efficiency requirements – Power Rating × 0.48 + 0.14<br>• Power Rating 1-49W: Efficiency requirements ≥[0.0626 × Ln(Power Rating)]+0.622<br>• Power Rating 50-250W: Efficiency requirements ≥87% |
| | **Output Voltage < 6V**<br>Power rating 0-49 watts: ≤ 0.3W of no load power<br>Power rating: 50-250 watts: ≤ 0.5W of no load power | • Power Rating ≤ 1W: Efficiency requirements – Power Rating × 0.497 + 0.067<br>• Power Rating 1-49W: Efficiency requirements ≥[0.075 × Ln(Power Rating)]+0.561<br>• Power Rating 50-250W: Efficiency requirements ≥86% |
| VI | **Output Voltage > 6V**<br>Power rating 1-49 watts: ≤ 0.1W of no load power<br>Power rating: 49-250 watt.s: ≤ 0.21W of no load power<br>Power rating: > 250 watts: ≤ 0.5W of no load power | • Power Rating 1-49W: Efficiency requirements ≥[0.071 × Ln(Power Rating)]-0.0014 × Power Rating +0.67<br>• Power Rating 49-250W: Efficiency requirements ≥88%<br>• Power Rating > 250W: Efficiency requirement > 87.5% |
| | **Output Voltage < 6V**<br>Power rating 1-49 watts: ≤ 0.1W of no load power<br>Power rating: 49-250 watts: ≤ 0.21W of no load power<br>Power rating: > 250 watts: ≤ 0.5W of no load power | • Power Rating 1-49W: Efficiency requirements ≥[0.0834 × Ln(Power Rating)]-0.0014 × Power Rating +0.609<br>• Power Rating 49-250W: Efficiency requirements ≥87%<br>• Power Rating > 250W: Efficiency requirement > 87.5% |

drop, effectively protecting the zener diode and any downstream electronics. The end result is an increase in the zener diode's power handling capability [15].

### 3.2.6 Power Supply Connectors

IEEE Std. 1725-2011 provides a list of requirements for AC adapters used in portable consumer electronic devices. In particular, the standard lists items that should be considered specifically for the power supply's input and output connectors [16]. These items include the following:

- The connector design should be such that it inherently prevents a reverse polarity connection of the AC adapter to the device that it powers.
- The AC adapter and device connectors should mate appropriately and be capable of good electrical contact through their useful lives (see Section 3.4.1 for a further discussion on this subject).
- The AC adapter connector's power and ground pins should be separated sufficiently to reduce the probability of a short-circuit condition.
- Connectors used on the output side must be capable of not only supplying the normal load current, but be able to handle short-term high current conditions during a fault in the external load. If the connector is not designed for this high transient current condition, the connector may be damaged by this event.
- The effect of high transient fault currents may be overlooked in output connectors used in adapters. The high fault current is not only due to the reduced load resistance, but also due to the discharge of the electrolytic filter capacitors. During a high current fault condition on the output, the transient current can exceed tens of amps even for relatively small adapters. The connector selection should include a thorough analysis of the magnitude of the fault current and the corresponding effect on the contact surfaces [17].

- The AC adapter's connector should be able to withstand multiple insertions and removals while maintaining the contact resistance within the maximum acceptable contact resistance specification over its expected lifetime.
  - During insertion, a connector is plugged into its mating surface. This requires an expenditure of force, the magnitude of which depends on several factors including the number of contact pins, individual pin size, pin construction and geometry, and the material used for plating the individual pins.
  - During extraction, a negative force is expended. It is desirable to have minimal insertion force per contact pin to allow for easy insertion without mechanical damage to the contact pins and the connecting cables. Conversely, it is desirable to have a high extraction force to ensure that the connector pins have low contact resistance, even under mechanical stress, vibration, and extended mechanical insertion/extraction cycles [17].

## 3.3 Adapter Specifications

The AC adapter specifications should be evaluated once the AC adapter has been selected for use in an application to ensure that it is adequately rated for the battery system and will not result in a shock, burn or fire hazard, or initiate a cascading failure that can cause a thermal runaway of the Li-ion cell. At a minimum, the parameters listed in Table 3.5 should be obtained from the AC adapter specifications or other documentation supplied by the adapter manufacturer. These specifications should then be matched with the specifications of the battery system's charger and protection circuit to ensure that the adapter is suitable for the battery system.

## 3.4 Power Supply Construction and Assembly Issues

AC adapters have been recalled because they can overheat or expose high voltage areas that increase the risk of an electrical shock hazard or lead to a fire. When evaluating an AC adapter to de-

**Table 3.5**
Typical USB Power Supply Specifications

| Requirement | Typical Set-Point |
|---|---|
| AC input voltage range | 90 to 264 VAC |
| Frequency | 47 to 63 Hz |
| No load power consumption | < 0.1W |
| Efficiency | > 81.9% |
| Output voltage | 5 V ± 250 mV |
| Output current | 2 A |
| Overcurrent protection | 2.5 A (autorecovery) |
| Short-circuit protection | No damage and autorecovery |
| Ripple | 150 mVp-p |
| Operating temperature range | 0°C to 45°C |
| Enclosure temperature | <50°C |
| Enclosure material | UL94 to V0 |
| Storage temperature | −25°C to 85°C |
| Storage humidity range | 0% to 95% RH |

This table provides the typical specifications for a USB type power supply (nominal 5V output and 2A output rating). All typical values listed in this table are for a USB power supply. These values will change for power supplies with different output voltage and current ratings.

termine its suitability for use in an application, it is important to perform a visual inspection and design review of the adapter's circuit to ensure that the adapter assembly/construction does not increase the risk of a failure that may lead to a shock, burn, or fire hazard. In addition to a visual inspection, the adapter should be thoroughly tested to understand its response to both normal operating conditions and foreseeable fault conditions.

### 3.4.1 Input Power Connection

The connection of the input AC power to the adapter's circuit board is an especially high risk area where an improper connection can lead to the risk of a propagating circuit board failure. Examples of potential issues are as follows:

- Spring-loaded connections are often used to provide an electrical connection between the input prongs of the adapter and its circuit board. The spring-loaded connections typically make physical contact with a pad/soldered area on the adapter's circuit board. Improper connection between the spring loaded connection and the circuit board due to issues such as poor alignment, inconsistent soldering on the circuit board, or inconsistent spring loaded connections can all lead to a relatively high resistive connection between the input spring loaded connection and the circuit board. This resistive connection can generate heat, pyrolize the circuit board, and lead to a propagating circuit board failure.

- Another often used configuration for connecting the AC input prongs to the power supply's circuit board involves the use of a connector lead and pin to connect the input AC prongs to the circuit board. The pin in this configuration is physically connected on one end to the connector lead and soldered to the adapter's circuit board on the other end. This configuration is especially susceptible to a fault at both the solder joint on the circuit board and the connector lead/pin connection. Furthermore, the probability of a failure of such a connection scheme increases with time. The physical movement of the power supply's plug and the mechanical stresses associated with this motion causes a degradation of the connection and can lead to a fracture of the solder joint. The intermittent contact at the fractured solder joint results in micro arcing events during the current interruption. This results in the thermal degradation of the insulating material in contact with both sides of the fracture, which eventually gets charred and becomes partially conductive. The charred insulating material provides a current path around the fracture and this current flow further heats and spreads the charred region until a conductive path is formed to an adjacent conductor, which then initiates the fault current. The risk associated with the connector lead and pin connection can be alleviated by the use of wires soldered to the adapter's AC input prongs and the circuit board.

### 3.4.2 Electrolytic Capacitors

Relatively large aluminum electrolytic capacitors with a relatively high voltage rating are often used in the adapter's input rectifier circuit. The failure modes of aluminum electrolytic capacitors are well documented and include [18]

- *Low-resistance failure:* A capacitor failure resulting in a low resistance is a consequence of a dielectric breakdown in the capacitor. Factors which may lead to this mode of failure include, but are not limited to, excessive voltage beyond its specified rating, excessive ripple current above its specified rating, and higher than specified ambient temperature.
- *Open-circuit*: A capacitor failure as an open circuit is usually due to a failure of the terminals of connections within the capacitor. Various factors such as corrosion, mechanical stresses in the form of shock, and/or vibration may contribute to the capacitor failing in this manner.
- *Capacitance loss:* This is generally due to a loss of electrolyte, which may be due to a leak in the capacitor seal over time. Excessive ripple current and high charge/discharge conditions coupled with a high ambient temperature may also lead to capacitance loss over a period of time.
- *Capacitor vent operation:* Capacitor destruction due to the operation of the capacitor venting mechanism is generally due to an internal pressure rise. This could be due an overvoltage condition, aging, or a capacitor short circuit due to either a direct short between the electrodes, a high ripple current above its rated value, or an insulation breakdown of the oxide layer on the aluminum foil.

Aluminum electrolytic capacitors are typically designed with a venting mechanism that releases the capacitor's internal pressure in a controlled manner. The venting of the electrolytic capacitor may result in the release of the capacitor's conductive electrolyte on to the adapter's circuit board, which poses the risk of resistive faults and a propagating circuit board failure [19]. For this reason, when selecting a mounting location for the electrolytic capacitor on to an adapter's circuit board, care should be taken to ensure that venting of the capacitor does not increase the risk of a circuit board failure.

### 3.4.3 Mechanical Damage and Access to High Voltages

Mechanical damage to the adapter's enclosure can lead to the exposure of the relatively high voltage components resulting in an electric shock hazard. Many AC adapter recalls have been attributed to enclosure mechanical damage induced electric shock hazards. Examples include

- Fujifilm recalled power adapter plugs sold with Fujifilm digital cameras in January 2018 because the power adapter wall plug could crack, break, or detach and remain in the wall, thus exposing live electrical contacts and posing a shock hazard [20];
- Power adapters sold with LectroFan sound machines were recalled in 2017 because the casing of the power adapter could break when plugged into an electrical outlet, exposing its metal prongs and resulting in an electrical shock hazard [21];
- Nvidia recalled plug heads for Shield AC wall adapters because the two-prong plug heads could break and pose a risk of electrical shock [22];
- Barnes and Nobles recalled power adapters in 2017 because the power adapter's casing could break when plugged into an electrical outlet, exposing its metal prongs and posing an electrical shock hazard [23].

When selecting an AC adapter for use in a battery application, the adapter should be tested to ensure that foreseeable mechanical abuse to the power supply enclosure does not result in exposing the adapter's high voltage components, which can lead to the risk of an electrical shock hazard.

### 3.4.4 Creepage and Clearance Distances

IEC 60950 defines clearance and creepage as follows [24]

- *Clearance:* The shortest path between two conductive parts, or between a conductive part and the bounding surface of the equipment, measured through air;

- *Creepage:* The shortest path between two conductive parts, or between a conductive part and the bounding surface[5] of the equipment, measured along the surface of the insulation.

Figure 3.10 demonstrates the concept of clearance and creepage distances in a PCB.

The AC adapter circuit should be designed to ensure that it meets the creepage and clearance requirements detailed in applicable standards. Identification of the clearance distance requires defining the installation category, the degree of pollution expected, the rated impulse withstand voltage, and the minimum required clearance specified in the applicable standard. Similarly, identification of the creepage distance requires defining the installation category, the degree of pollution expected, and the rated voltage [25].

The layout of any PCB traces should be designed in conformance with the guidance in industry standards, such as UL 60950-1 and IPC 2221. According to ANSI/UL 60950-1, a minimum clearance of 0.6 mm is required to achieve functional insulation performance for coated printed circuit boards with a working voltage between 200 and 250 Vrms or DC. The minimum clearance recommended for uncoated areas (such as exposed solder joints)

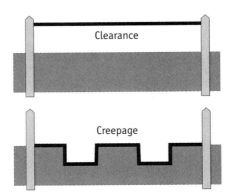

**Figure 3.10** Clearances and Creepage in a PCB.

---

5. Bounding surface: The outer surface of an electrical enclosure, considered as though metal foil were pressed into contact with accessible surfaces of insulating equipment. An example is a metal enclosure that the PCB is mounted into.

is 1.5 mm for voltage levels higher than 150 Vrms, or 210 Vpk/DC, and up to 300 Vrms or 420 Vpk/DC in primary circuits.

### 3.4.5 Soldering/Contaminants and Other Assembly Issues

Improper assembly of AC adapters can often be a cause of field failures, which may or may not lead to safety issues. Circuit board failures due to contaminants, premature component failures due to damage during the soldering process, or improper handling of the adapter during assembly are all common causes of field issues. These issues can lead to a relatively benign failure of the adapter where it simply stops functioning or it can result in a safety hazard if the contaminants lead to a propagating circuit board failure. Although it is a common expectation that an adapter will be assembled using industry standard methodology, that is not always the case. Often an inspection of several samples of an AC adapter can demonstrate whether the assembly process is well controlled and help identify any risks that should be mitigated before mass production. As an example, Figure 3.11 shows a portion of the circuit board of an AC adapter purchased by the authors for review. As can been seen in the figure, this adapter has a relatively high risk of failure in the field due to the presence of contaminants, inconsistent soldering, and improper mounting of the electrolytic

**Figure 3.11** Poor manufacturing controls can lead to AC adapters with elevated risk of field failures.

capacitor. This indicates that the assembly/soldering process used by the manufacturer for this adapter was not well controlled.

## 3.5 Testing AC Adapters

Once an AC adapter design is drafted and prototype samples are available, testing should be performed to characterize the operation of the adapter and identify any weaknesses that can lead to failures or other issues in the field. Testing should be focused on both characterizing the operation of the adapter under normal operating conditions and determining the adapter's response to foreseeable fault conditions.

### 3.5.1 Electrical Characterization Tests

Electrical characterization tests should focus on ensuring that all components on the AC adapter's circuit board operate within their rated specifications and are adequately derated. The following electrical characterization tests should be considered when evaluating candidate adapters:

- *Efficiency measurements:* As discussed in Section 3.2.4, AC adapters must comply with specific efficiency requirements that depend on the AC adapter's rating. Efficiency measurements should be performed to evaluate both the adapter's no load power consumption and its efficiency at the pertinent loading conditions to ensure that the adapter circuit has the ability to meet the mandatory efficiency requirements.
- *Component voltage/current characterization:* Measurements should be performed on all components used in the adapter circuit to ensure that the components are operated within their maximum voltage and current ratings with adequate margin. These measurements should be performed under different input voltage and output load conditions. These measurements can be used in combination with circuit simulation tools to determine the expected reliability of the adapter circuit.

- *Insulation resistance measurements:* The adapter circuit will include both a high voltage AC input and a lower voltage DC output. Insulation resistance measurements should be performed between the input(s) and output(s) of the adapter circuit to ensure that the adapter output is adequately insulated from the higher voltage input.
- *Hi-pot test:* The hi-pot test is essentially a test of the insulation surrounding the high voltage primary circuit in the adapter. A successful hi-pot test ensures that the insulation surrounding the primary circuit provides adequate protection against an electric shock hazard. The hi-pot test should be performed between both the adapter circuit's input and output, and also the adapter circuit's input and its enclosure.
- *Dynamic load test:* The dynamic load test is performed to investigate the adapter circuit's ability to regulate its output voltage under dynamic load conditions. The selection of the dynamic load conditions for the test is application-dependent.
- *Output short circuit:* The adapter's response to an output short-circuit should be characterized to ensure that this condition does not cause components on the adapter circuit to operate outside their rated specifications that can lead to a premature failure of the components.
- *Output overcurrent condition:* Adapter circuits should be designed with overcurrent protection that shuts the adapter output once the output current exceeds a predesigned threshold. Testing should be performed to ensure that the output overcurrent protection circuit reliably detects an output overcurrent condition and shuts the adapter output without resulting in power components being overstressed

### 3.5.2 Thermal Characterization Tests

The aim of the thermal characterization tests is to determine the operating temperatures of the components on the adapter circuit board under both normal and fault conditions. In addition, the thermal characterization tests should also characterize the temperatures of the adapter enclosure to ensure that the enclosure temperature does not rise to levels that may lead to a burn hazard. The following guidelines should be followed when performing the tests.

- The tests should be performed on fully assembled adapters to simulate real world operating conditions.
- The tests should also simulate restricted air flow conditions. It is not uncommon for adapters to be operated under blankets or other such areas where they are insulated from the ambient environment. The component temperatures under such restricted air flow conditions can be substantially higher that the component operating temperatures under normal ambient conditions.
- A thermal imaging camera can be used to identify hot spots on the adapter's enclosure to ensure that the worst case enclosure temperatures are identified during the tests.

### 3.5.3 Mechanical Abuse Tests

Access to the adapter's high voltage input circuit due to a compromise of its enclosure can lead to an electrocution hazard. For this reason it is important to characterize the response of the adapter to foreseeable mechanical abuse conditions. These conditions should simulate the adapter's response to impact, drop, shock, vibration, and elevated mechanical load conditions. When performing the mechanical abuse tests, it is important to also consider different ambient temperatures as the response of the adapter's (typically plastic) enclosure may be different at different operating conditions.

### 3.5.4 Single Points of Failure Tests

An adapter circuit may include a large number of components and often it may not be practical to characterize the response of the adapter to a failure of all components. When selecting components for the single point of failure tests, it is important to focus on components whose failure increases the risk of

- A propagating circuit board fault as they are in the high-power path (e.g., input filter capacitors, and switching transistors);
- An electrical shock hazard as they may compromise the adapter's insulation mechanism (e.g., Y-capacitors[6]);
- Components operating at elevated temperatures as they are part of the power supply's temperature monitoring circuit (e.g., NTCs[7]);
- A power supply output overvoltage condition as they are part of the power supply's voltage regulation circuit (e.g., optocouplers or voltage divider circuits connected to the pulse width modulation control IC).

## 3.6 DC-DC Converter Circuits

DC-DC converter circuits (DC-DC power supplies) can be used to either step down a DC voltage (typically called buck converters), or step up a DC voltage (typically called boost converters). The most common use of DC-DC power supplies in portable consumer electronic devices is in automobiles. Today, most automobiles are equipped with USB outlets that provide charging stations for consumer electronic devices. DC-DC power supplies in automobiles are used to step down the nominally +12V automobile battery voltage to a regulated 5 Vdc output for use by consumer electronic devices. Car chargers for portable consumer electronic devices are another very similar application of DC-DC power supplies.

Most DC-DC power supplies used for portable consumer electronic devices are switch-mode power supplies and are based on operating principals similar to AC-DC switched mode power supplies. The requirements and risks of these power supplies are also similar to AC-DC power supplies. When used in an automotive

---

6. Y-capacitors are commonly used for EMI filtering applications where they are directly connected across the AC power line. The definition of a Y-capacitor application is "where damage to the capacitor may involve the danger of electric shock." [26]
7. A negative temperature coefficient (NTC) thermistor is a device that exhibits a decrease in electrical resistance with increasing temperature. NTC thermistors are often used to monitor temperatures in a variety of applications.

environment, DC-DC power supplies have some additional requirements. These include:

- The need to operate in ambient temperatures that can be significantly higher than in a typical home. As an example, Figure 3.12 shows some of the measured temperatures in an automobile on a hot summer day in Phoenix, Arizona. As can be seen in the figure, the ambient temperatures in a vehicle can exceed 70°C with the temperature on the dashboard getting above 95°C. This results in the need for DC-DC power supplies to monitor ambient temperatures to ensure that the device protects its components and prevents them from operating outside their rated temperature. Additionally, for DC-DC power supplies that are connected to a car battery, any failure in the power supply resulting in a short-circuit of the automobile's high current outlet can lead to a fire hazard unless it is adequately protected.

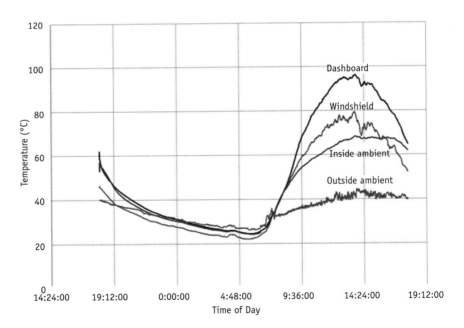

**Figure 3.12** Temperatures measured in an automobile on a summer day in Phoenix, Arizona.

- The need to continue functioning with a constantly varying input power supply that generates numerous electrical transients on a continuous basis. The +12V (or 24V in larger vehicles) power supply system in an automobile generates transients as different loads in the automobile turn on and off. The DC-DC power supply must be able to absorb this transient energy without failure and continue functioning in this environment. The international standard ISO 7637 [27] describes a set of tests that can be used to evaluate whether a DC-DC power supply will function properly when connected to the automobile's power system.

# References

[1] Pressman, A., *Switching Power Supply Design*, Second Edition, New York: McGraw-Hill, 2014.

[2] AND9009/D, Types of Electrical Overstress Protection, ON Semiconductor.

[3] Designing with Thermally Protected TMOV® Varistors in SPD and AC Line Applications, Application Note EC635, Littelfuse.

[4] Blanchard, R., and N. K. Medora, et al., "Failure Analysis of Printed Wiring Assemblies," *Electronic Failure Analysis Handbook*, New York: McGraw Hill Publishing Company, 1999, Chapter 14, pp. 14.1–14.27.

[5] Slee, D., J. Stepan, W. Wei, and J. Swart, "Introduction to Printed Circuit Board Failures," *IEEE Symposium on Product Compliance Engineering*, Toronto, Ontario, Canada, 2009, pp. 1–8.

[6] https://cpsc.gov/Recalls/2008/power-adapters-used-with-notebook-computers-recalled-by-battery-biz-due-to-burn-hazard.

[7] Nute, R., "Principles of Thermally-Caused [sic] Injury" *The Product Safety Engineering Newsletter*, Sept. 6–12, 2007.

[8] Raj, P.K., "Hazardous Heat," *NFPA Journal*, Sept./Oct. 75–79, 2006.

[9] BS EN 13202:2000. 2000. Ergonomics of the Thermal Environment - Temperatures of Touchable Hot Surfaces–Guidance for Establishing Surface Temperature Limit Values in Production Standards with the Aid of EN563, Annex B, Table B.1. Available (for purchase) from SAI Global: http://infostore.saiglobal.com/store/details.aspx?ProductID=878308.

[10] UL. 2007. UL 60950-1, Second Edition. Information Technology Equipment–Safety–Part 1: General Requirements, Table 4C–Touch Temperature Limits. See http://ulstandards.ul.com/standard/?id=60950-1_1.

[11] ASTM. 2003. Standard Practice for Determination of Skin Contact Temperautre from Heated Surfaces Using a Mathematical Model and Thermesthesiometer. (n.d.). ASTM C 1057–03, American Society for Testing and Materials.

[12] ASTM. 2003. Standard Guide for Heated System Surface Conditions that Produce Contact Burn Injuries. (n.d.). ASTM C 1055–03, American Society for Testing and Materials.

[13] http://www.littelfuse.com/~/media/electronics/application_guides/littelfuse_fuseology_application_guide.pdf.pdf.

[14] https://www.gpo.gov/fdsys/pkg/USCODE-2010-title42/html/USCODE-2010-title42-chap77-subchapIII-partA-sec6291.htm, accessed May 13, 2018.

[15] PolyZen Protection Device for USB Applications, Datasheet, Tyco Electronics.

[16] IEEE Std 1725-2011, IEEE Standard for Rechargeable Batteries for Cellular Telephones, Section 8.

[17] Medora, N.K., "Connection Technology," *Electronic Failure Analysis Handbook*, New York: McGraw Hill Publishing Company, 1999, Ch. 17, pp. 17.1–17.69.

[18] Arora, A., et al., "Failures of Electrical/Electronic Components: Selected Case Studies," *IEEE PSES*, Longmont, CO, Oct. 22–24, 2007.

[19] Kaiser, C. J., *Aluminum Electrolytic Capacitors, The Capacitor Handbook*, Olathe, KS: C J Publishing, 1995, Ch. 4.

[20] https://www.cpsc.gov/Recalls/2018/fujifilm-recalls-power-adapter-wall-plugs-sold-with-digital-cameras-due-to-shock-hazard, accessed May 17, 2018.

[21] https://www.cpsc.gov/Recalls/2017/power-adapters-sold-with-lectrofan-sound-machines-recalled-by-asti-recall-alert, accessed May 17, 2018.

[22] https://www.nvidia.com/en-us/shield/support/adapterrecall/, accessed May 17, 2018.

[23] http://wemakeitsafer.com/Power-Adapter-Recall-888672-118435, Accessed May 17, 2018.

[24] Optimum Design Associates, "Clearances and Creepage Rules for PCB Assembly," http://blog.optimumdesign.com/clearance-and-creepage-rules-for-pcb-assembly.

[25] Harting, "Creepage and Clearance Distances," http://files.may-kg.com/td_334_1.en.pdf, accessed May 17, 2018.

[26] https://www.okaya.com/noise-products/capacitors/capacitor-faq/.

[27] International Standard, ISO 7637-2, Road Vehicles–Electrical Disturbances from Conduction and Coupling.

# 4

# Li-Ion Battery Pack Charge Circuits

The specifications that the charge circuit will need to follow will depend on the cell chemistry used for the application. This chapter will use Li-ion cells with a lithium cobalt dioxide cathode and a graphite anode as an example for the discussion. In addition, the chapter will focus on Li-ion battery charger[1] circuits for portable consumer electronic devices. While the design and requirements for charger circuits are similar regardless of the application, the implementation of charger functionality can vary significantly between applications. As an example, charger circuits in portable consumer electronic devices are often located on the device's circuit. A DC power source (AC adapter) connects to the device and is conditioned by the charger circuit in the device. On the other hand, applications, such as electric scooters, may contain the charging circuitry as part of the battery protection circuit. A DC power source in these applications interfaces directly with the battery.

---

1. The term charge and charger circuits is used interchangeably in this book.

## 4.1 Charge Profile

The charger circuit regulates and conditions the DC output voltage from the AC adapter connected to the device and provides a constant-current constant-voltage (CCCV) charge profile to the Li-ion battery. Figure 4.1 shows the typical charge profile of a single-cell Li-ion battery (the profile shown in the figure is a typical charge profile for a single Li-ion cell with a lithium cobalt dioxide cathode and a graphite anode). For a single-cell battery, a Li-ion battery's charge algorithm will typically consist of the following stages:

- *Overdischarged battery detection (also often referred to as the 0V charging mode):* Smart Li-ion battery chargers are often designed to disable charging of a battery if the voltage of any cell in the battery is below approximately 2.0V (a range of 1.5V to 2.0V is commonly used. It is not uncommon for this functionality to be implemented in the battery protection circuit).

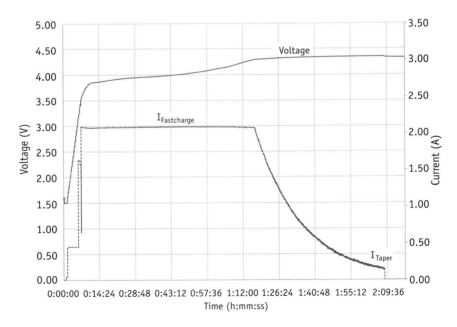

**Figure 4.1** Typical charge profile for a Li-ion battery.

- *Precharge*: Li-ion batteries are typically charged at a reduced charge current magnitude if the voltage of any cell in the battery pack is below approximately 3.0V. A charge current rate in the vicinity of $0.1C^2$ is often used as the precharge current.
- *Fast-charge*: The charger transitions from the precharge to the fast charge mode of operation when the voltages of all the cells in the battery pack rise above 3.0V (a voltage threshold in the range of 2.7V to 3.0V is often used for the transition from the precharge to the fast charge mode.). In the fast charge mode, the battery is initially charged at a constant current until the battery voltage reaches 4.2V. At this point, the charger transitions to the constant voltage mode. The charge current continues to flow until it drops to a pre-set value where the charging is terminated (this termination current is often specified in the cell datasheet. A value of 0.1C is often used as the termination current value).

The charge circuit constantly monitors the battery voltage and reinitiates charging when the battery voltage drops below a preset value (typically 4.1V).

Multicell batteries typically follow a similar charge profile with both a constant current and a constant voltage mode during the charge cycle. The open-circuit output voltage applied by the charger to charge the battery is a function of the number of Li-ion cells connected in series and is a multiple of 4.2V (some Li-ion cells are specified for charge voltages up to 4.35V). For example, the charger outputs a voltage of 8.4V (4.2V × 2) for battery packs with two series connected cells and 12.6V (4.2V × 3) for battery packs with three series connected cells (this value does not change if multiple cells are connected in parallel ). Similar to the termination condition for single-cell Li-ion batteries, the charge current for multicell Li-ion batteries is terminated once the current drops to a preset value during the constant voltage phase of the charge cycle.

---

2. The C-rate is used to express charge and discharge rates. The symbol "C" is followed by a number, preceded by a number or decimal portion of an integer, or divided by a number. When used to represent the charge and/or discharge current, the current is referenced to a specific cell/battery capacity. For example, if a cell rated for 2 Ah is being discharged at 500 mA, its discharge rate is 0.25C or C/4.

## 4.2 Safety Timers

Smart Li-ion battery chargers are designed with safety timers that terminate the charge current if the battery is not charged within a preset time interval. Safety timers are implemented for both the precharge mode of operation and when the charger is operating in the fast charge mode. Due to the steep voltage curve of a Li-ion cell when it is fully discharged, it takes only a few minutes for a Li-ion cell to charge from 2.0V to 3.0V even when it is charged at a reduced charge current (around 0.1C). If a Li-ion cell has not charged above 3.0V for a prolonged period of time (approximately 30 minutes), this could be a sign of a fault within the cell that is preventing the cell's voltage from increasing and the cell from charging. In this scenario, it is important to not continuously charge the faulty Li-ion cell as the charge current will likely be converted into heat by the cell. For this reason, smart chargers are designed with a precharge timer that terminates the precharge current if the charger does not transition from the precharge to the fast charge mode of operation within a certain time interval. A precharge timer setting of 30 minutes is commonly used for smart chargers in portable consumer electronic devices.

Similar to the need for a precharge safety timer, a fast charge safety timer is also implemented in smart chargers to reduce the risk of a battery failure due to a faulty cell. As is the case in the precharge mode, a longer than normal charge time during the fast charge mode of operation for a battery could be a sign of a resistive fault within the cell that is preventing it from being fully charged. In this scenario, continuing to supply current to the battery can lead to the conversion of this current into heat, which may increase the probability of a cell failure. It is common for smart chargers in portable consumer electronic devices to be designed with a fast charge safety timer that terminates the charge current if the charge current has been flowing for 130% to 150% of the time that it would take to fully charge a battery. As an example, Figure 4.2 shows the operation of the safety timer in a single-cell battery charger circuit. As can be seen in Figure 4.2, the charge current is terminated by the charger circuit after approximately 3 hours as the battery does not fully charge during this period.

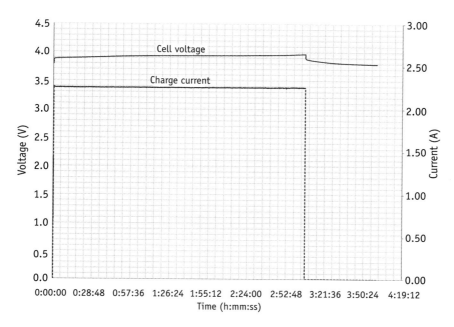

**Figure 4.2** Operation of the fast charge safety timer in a single-cell battery charger circuit.

## 4.3 Battery Temperature Monitoring

Smart chargers in portable consumer electronic devices are also often designed to monitor the Li-ion battery temperature and only charge the battery if the temperature is within the allowable charging temperature range for the battery. Temperature sensors (such as thermistors and NTCs)[3] are often located in the battery and provide battery temperature information to the charger circuit.

### 4.3.1 Charging Li-Ion Cells at Low Temperatures

Li-ion cells are assembled in a discharged state. The first charge of the cell results in the formation of an electrolyte decomposition layer on the surface of the negative electrode of the cell. This layer is called the solid electrolyte interface layer (commonly referred to as the SEI layer). The SEI layer maintains a protective barrier between the cell's reactive negative electrode and the electrolyte.

---

3. These are discussed further in Chapter 5.

The SEI layer is porous enough to allow the passage of lithium ions. However, the SEI layer limits the discharge rates and restricts the temperature over which a Li-ion cell may be charged. At low temperatures, the rate of lithium ion transport through the SEI layer is hindered. For this reason, charging a Li-ion cell at low temperatures can result in lithium plating at the SEI/electrolyte interface layer. Lithium plating occurs when the lithium ions deposit as metallic lithium on the negative electrode surface during charge instead of intercalating into the negative electrode's active material. Charging a cell at low temperatures is not the only cause of lithium plating in Li-ion cells. Lithium plating in a cell can occur due to a variety of other reasons, such as normal degradation of the negative electrode due to aging, a fast charge rate, overcharging of the cell, or fast charge rates for cells that are at a low state of charge (i.e., the absence of a precharge mode of operation). Figure 4.3 shows a negative electrode from a Li-ion cell that exhibits signs of lithium plating. The plated lithium on the negative electrode forms lithium hydroxide when exposed to air and can be seen as the white foam type substance in this figure.

The following can occur to the plated lithium in the cell:

**Figure 4.3** A negative electrode from a Li-ion cell showing signs of lithium plating. The plated lithium forms lithium hydroxide when exposed to air and can be seen as the white foam type substance in this picture.

- Redissolution during a subsequent discharge cycle through a process called electrochemical stripping.
- The formation of dead lithium deposits.
- The reaction of the plated lithium with the electrolyte to form the SEI layer. This reaction can lead to a reduced capability of the cell in handling higher charge and discharge current due to an increase in the cell's impedance. It also enhances the likelihood of subsequent lithium plating and of localized overdischarge in the cell.
- The plated lithium can give rise to dendrites that can cause shorting within the cell. A mat of dendrites and dead lithium can increase the likelihood that a minor short-circuit within a cell will lead to a cell thermal runaway condition.

### 4.3.2 Charging Li-Ion Cells at High Temperatures

Charging Li-ion cells at high temperatures (typically above 45°C) increases the risk of gas generation within the cell, which can lead to cell swelling, nuisance operation of the pressure triggered protective devices in the cells (such as the current interrupt devices in cylindrical cells), or thermal runaway due to mechanical disturbance of windings or layers within the cells. Charging Li-ion cells at elevated temperatures and charge current rates can degrade the performance and life of the cells.

## 4.4 Battery ID

It is not uncommon for both single and multicell battery chargers to incorporate a means of identifying the battery before providing a charge voltage and current to the battery. This capability is more commonly implemented in multicell battery chargers although single-cell battery chargers also often use some means of determining the identity of the battery connected to the charger. IEEE 1725-2011 [1] states the following regarding battery pack identification:

An identification scheme shall be employed so that the host device can identify the type of battery pack present in the system. At a minimum, the identification scheme shall communicate or indicate the maximum charge voltage. If the battery pack cannot be properly identified, the charge shall be terminated. Discharge may be terminated. In addition to an identification scheme, a mechanical interlock that will impede the installation of an improper pack should be incorporated as well.

Single-cell battery chargers often use simple checks to identify the battery. This often includes checking to see if a specific resistor is installed in a specific location on the battery protection circuit prior to initiating charging. Multicell battery chargers on the other hand often employ relatively more sophisticated communication links to get information from the battery. The I2C communication scheme (SMBus protocol) is commonly used by battery packs and chargers to communicate with each other. The I2C bus is a standard bidirectional interface that uses a controller, known as the master, to communicate with slave devices. The I2C is a synchronous, multimaster, multislave, packet-switched, single-ended, serial computer bus that is widely used for attaching lower-speed peripheral ICs to processors and microcontrollers in short-distance, intraboard communications. The physical I2C interface consists of the serial clock (SCL) and serial data (SDA) lines [2].

## 4.5 Charger Specifications and Requirements

The charger circuit specifications should be carefully evaluated once a charger design has been selected to ensure that the charger circuit is designed to prevent charging a cell outside its rated specifications under all foreseeable normal operating and fault conditions. At a minimum, the parameters listed in Table 4.1 should be obtained from the charger specifications or other documentation supplied by the charger manufacturer. Table 4.1 provides a summary of some common specifications of both single-cell and multicell battery charger circuits.

**Table 4.1**
Typical Charger Specifications

| Requirement | Typical Set Point Single-Cell Chargers | Multicell Chargers |
|---|---|---|
| Open circuit output voltage | ~4.2V | ~4.2 × (# of series connected cells) |
| Charge profile | CCCV | CCCV |
| Charge termination | Current drops to ~C/20 | Current drops to ~C/20[1] |
| Charge initiation | Battery voltage < 4.1V | (Any parallel cell combination voltage < 4.1V) & (battery pack voltage < 4.1V × (# of series connected cells)) |
| Discharge termination[2] | Battery voltage < 3V | Typically not incorporated in chargers for multicell Li-ion battery packs |
| Precharge initiation | Battery voltage < 2.7V | Any parallel cell combination voltage < 2.7V |
| Charge current | 0.5C to 0.7C | 0.5C to 0.7C |
| Precharge current | ~0.1C | ~0.1C |
| Zero volt charge | No charge current if battery voltage < (1.7V – 2.0V) | No charge current if any parallel cell combination voltage < (1.7V – 2.0V) |
| Fast charge timeout | 1.3–1.5 times the charge time | 1.3–1.5 times the charge time |
| Precharge timeout | 30 min | 30 min |
| Input overvoltage | > 12V | Depends on the number of cells/cell groups connected in series |
| Output short circuit | Terminates charge current on detecting an output short-circuit condition | Terminates charge current on detecting an output short-circuit condition |
| Charge status | Provides information on charging status | Provides information on charging status |
| State of charge[3] | Updates, stores, and provides state of charge information, which may be displayed to a user | Updates, stores, and provides state of charge information, which may be displayed to a user |
| Charge temperature | 0°C to 45°C | 0°C to 45°C[4] |
| Battery identification[5] | Only charges batteries designed and manufactured for use with the charger | Only charges batteries designed and manufactured for use with the charger |
| Battery communications[6] | Only provides a charge current during active communications with the battery | SMBus communications between charger and battery pack |

1. The total C rating of the battery pack is used to determine the charge cut-off current.
2. This feature is only relevant for systems where the discharge current also flows through the charge circuit.
3. State of charge is also commonly monitored by the battery protection circuit.
4. For multicell battery packs, temperature protection is typically included as part of the battery pack's protection electronics.
5. This is typically only implemented in devices with user replaceable batteries.
6. This feature is only relevant for systems where the charger circuit communicates with the battery pack.

## 4.6 Charger Circuit Construction and Assembly Issues

A visual inspection of the charger should be performed to ensure that the charger circuit layout reduces the probability of a propagating circuit board failure[4] (for example, adequate spacing between PCB traces at different voltages). The trace layout should be designed in conformance with the guidance within standards such as UL 60950-1 and IPC 2221. In addition, the general quality of the manufactured charger circuit should be reviewed on multiple samples to ensure that the charger construction is consistent and will not lead to reliability issues in the field. At a minimum, the following items should be reviewed and visually inspected:

- *Output connector*: Typically, the charger output connector will consist of multiple pins. These pins will include power and communication pins. The connector's design should be such that the positive and negative terminals are at the two opposite ends of the connector (i.e., the positive and negative terminals are separated as much as possible at the connector).
- *Temperature monitoring:* Charger circuit designs that do not rely on temperature information from the battery pack but contain their own temperature monitoring circuits should have the monitoring circuits physically located as close as possible to the battery pack. The effectiveness of this temperature monitoring circuit should be determined specifically on multicell battery packs to ensure that the temperature sensor can provide an accurate reading of the ambient condition around the cells in the battery pack. Ideally, temperature monitoring circuits should be incorporated inside battery packs as part of the PCM circuit to ensure accurate monitoring of ambient temperatures around the cells.
- *Solder quality:* The quality of the solder connections (specifically in and around the output connector) should be inspected for cold solder joints, solder splatter, cracks, and/or contamination. This inspection should be performed on multiple charger samples to evaluate the consistency of the soldering process.

---

4. See Chapter 3 for a discussion on propagating circuit board failures.

- *Flux residue:* The charger printed circuit board should be inspected for excessive flux residue. Again, this inspection should be performed on multiple charger samples to evaluate the consistency of the soldering process. Excessive flux residue, even with no-clean flux, can lead to an accumulation of contamination during processing and in service. Such contamination can result in short-circuit paths between components where excessive solder flux is present.

## 4.7 Testing Charger Circuits

Once the design and assembly of the charger circuit is completed, it is important to perform testing on samples to ensure that the charger circuit operates as intended and provides the desired protection under all foreseeable fault conditions. Different charger circuit designs and configurations demand different protocols to be followed when evaluating the charger circuit. The type of tests performed will depend on factors such as the charger circuit design, the size of the battery, and the application that the charger is used in. The following categories of electrical tests are often performed when evaluating charger circuits regardless of the size of the battery that is being charged.

- *Charge profile verification:* It is important for the charger circuit to provide an appropriate charge voltage and current to the battery. In addition, the charger circuit should prevent the battery from being trickle charged. Typically the following are desirable characteristics in charge profile:
    - The charger should use a constant current-constant voltage profile to charge the battery.
    - The charger should provide an appropriate charge current and voltage to charge the battery. The charge voltage and current should not exceed the maximum values listed in the battery specifications.

    As an example, Figure 4.4 shows an example charge profile for a multicell battery (four cells connected in series). As

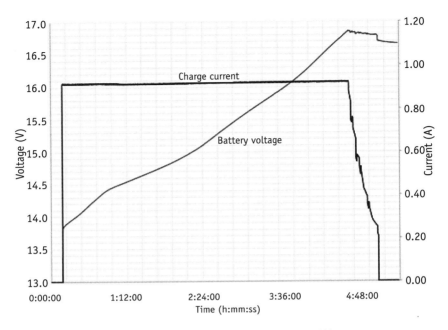

**Figure 4.4** Example charge profile for a multicell battery.

can be seen in the figure, the charger initially operates in a constant current mode until the battery voltage rises to approximately 16.8V at which point the charge transitions to the constant voltage mode of operation. The charge current terminates once it drops to approximately 220 mA.

- *Fast charge timeout characterization:* As discussed in Section 4.2, it is important for the charger to not continue to supply charge current to a failed battery, as that increases the probability of a battery failure. For this reason, charger circuits should be designed with a safety timer that terminates the charge current after a certain predesigned period of time. Fast charge timeout characterization tests should be performed to investigate the ability of the charger to timeout if unable to charge a battery within a preset time period. In this test, the voltage of the battery being charged is clamped at a fixed value to prevent the battery from fully charging during the test. For single-cell batteries, the voltage of the cell is typically clamped at 3.7V using an electronic load in the constant voltage mode connected directly across the cell terminals.

For multicell battery packs, the electronic load is connected directly across the cell stack and set to clamp the stack voltage at 3.7V × (number of series cell combinations). It is ideal for the charger to time out within 130% to 150% of the normal charge time.

- *Precharge characterization:* It is important to ensure that the charger circuit is designed to provide a lower magnitude charge current (precharge current) to an overdischarged battery. Precharge characterization tests should be performed to determine if the precharge current setting is appropriate for the battery. For operating the charger in the precharge mode, the battery being tested is over-discharged for this test. For single-cell battery packs, this is done by reducing the voltage of the cell to approximately 2V using an electronic load in the constant voltage mode connected directly across the cell terminals. For multicell battery packs the voltage of the cell stack is reduced to approximately 2V × number of series cell combinations (due to the steep drop in the cell voltage curve at low states of charge, a relatively large cell imbalance may be observed when the series cell combination are at voltages below 3V. In the event of a large cell imbalance, the electronic load should be set to ensure that the voltage of the lowest series cell combination is approximately 2V at the start of the test). The battery is then connected to the charging circuit to evaluate how the charger charges the overdischarged battery. Ideally, the charger should provide a precharge current (generally equal to or less than a C/10 rate) to precondition the over-discharged battery pack and should transition from the precharge to the fast charge mode when the cell voltage rises to approximately 3V for single-cell battery packs, or when each individual series cell combination rises to approximately 3V for multicell battery packs. As an example, Figure 4.5 shows the operation of a charger for a 4 Ah single-cell Li-ion battery in the precharge mode and its transition from the precharge to the fast charge mode when the battery voltage rises above approximately 3.0V.

- *Allowable charge temperature characterization:* This characterization is only performed if the charger circuit monitors the

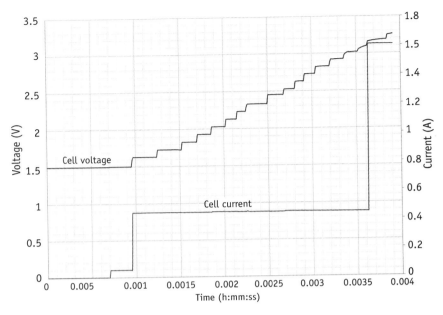

**Figure 4.5** Precharge mode and precharge to fast charge characterization in a single-cell battery system.

battery pack temperature and uses this information to determine when to charge the battery (for multicell batteries, it is not uncommon for the temperature to be monitored by the battery protection circuit and for the battery protection circuit to determine when to allow charge current to the battery). The aim of this characterization is to verify that the battery protection circuit/charge circuit will prevent charging outside the allowable charge temperature range for the cells. As an example, Figure 4.6 shows the results of an allowable charge characterization test performed on a multicell battery charger. As can be seen in the figure, the charger only charges the battery pack for cell temperatures from approximately 0°C to 45°C.

- *Charger output short-circuit:* This characterization simulates a failure at the output of the charger circuit that exposes the charger to a short-circuit condition. Ideally, the charger should be able to detect an output short-circuit condition and terminate charging on detecting the short circuit. This is nec-

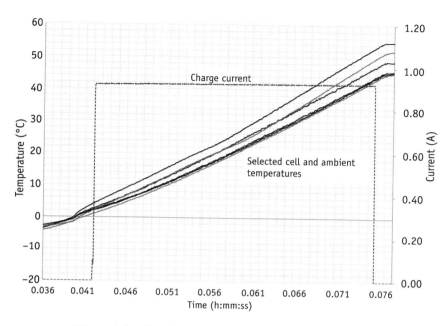

**Figure 4.6** Allowable charge temperature characterizations.

essary so that the charger does not provide any energy to a failed component on the battery protection circuit or a cell with an internal short-circuit.

- *Input overvoltage condition*: The use of an incorrect input power source (such as an incorrect AC adapter) can result in a cascading failure of the battery system, eventually leading to a cell overcharge condition if the components are not rated to handle an overvoltage condition. For this reason, it is important to characterize the response of the charger circuit to an input overvoltage condition. The aim of this characterization is to determine the voltage at which components start to fail in the battery system. This will allow for the identification of a need for a dedicated overvoltage protection circuit at the charger input if the charger circuit cannot handle foreseeable input over-voltage conditions.

- *Communication interruption:* For charger circuits that communicate with the battery, and get charge voltage and current information from the battery, it is important for the charger circuit to detect an interruption or failure in this communication and terminate the charge current if that happens. The aim of this characterization is to evaluate the response of the charge circuit to a failure of the communications link between the charger and the battery.

- *Cell imbalance detection:* This characterization is only necessary for multicell batteries. In multicell batteries, it is not uncommon for the cell imbalance detection to be performed by the battery protection circuit. The aim of the cell imbalance detection is to permanently prevent the battery pack from being charged if the cell imbalance between the various series connected cells exceeds a pre-set threshold. This threshold is typically (though not always) set at 0.5V. The development of a large cell imbalance in a battery signifies that a battery has reached its end of life, or that one or more cells in the battery have degraded or developed an issue. For this reason, battery protection circuits or battery charger circuits are designed to essentially disable the battery pack by not providing it a charge current if a large cell imbalance develops. The cell imbalance detection is typically performed only during the charge cycle and also once the cell voltage rises above approximately 3.7V. This is done to prevent false positives associated with the steep voltage curve for Li-ion cells specifically below 3.5V.

## 4.8 Wireless Charger Circuits

Wireless chargers rely on the principle of induction. Wireless chargers are especially popular for portable devices, such as cell phones, where the charging cable is connected to a charging base or a charging station instead of the phone and the phone is simply placed on the charging station. Wireless chargers typically provide the advantage of enhanced reliability as the charging cable does not have to be repeatedly plugged into the device, thus preventing mechanical damage both at the device and cable ends. Wireless chargers generally follow a standard, such as the Qi standard, which defines wireless power transfer using inductive charging over small distances. Any device that is compatible with these standards can be charged using the wireless charger. The main components of a wireless battery charging system are as shown in Figure 4.7.

A device utilizing a wireless charger will typically consist of [3]

- A wireless charging transmitter that is typically powered by an AC adapter.
- A transmitter switching and resonant circuit.
- One or more transmitter coils used to transfer power to the receiver electromagnetically (Figure 4.8 shows an example of such a coil in a commercial wireless charger).
- A receiver which consists of a similar coil.
- A receiver circuit that consists of a battery charger circuit, which will take the input power from the wireless charging base station and provide the correct voltage and current to charge the battery. In addition, the battery charger circuit will contain all the other features discussed in previous sections of this chapter.

Although the means of transferring power from the wall to the device is different, all the design considerations that are taken into account as discussed in this chapter are also applicable to a wireless charging system. Any fault that may result in the failure of the wireless charging system should not propagate to the battery

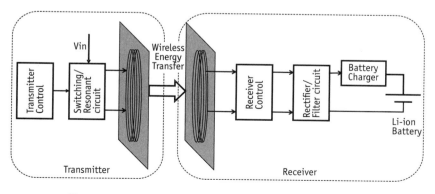

**Figure 4.7** Components of wireless battery charging system.

**Figure 4.8** Typical commercial wireless charger.

system. Due to the nature of induction, charging wirelessly is comparatively a slower process than charging directly using cables. There are also a few challenges while designing an efficient wireless charger due to the additional coil circuitry within the charger. The charger should be designed such that it is shielded from external noise and EMI that may affect the coil operation. The charger also has a limitation with respect to the material that can be used for its enclosure as the thickness, and the material type should be such that it causes minimum hindrance in power transfer between the transmitter and the receiver.

# References

[1] IEEE 1725-2011: IEEE Standard for Rechargeable Batteries for Cellular Telephones.

[2] http://www.ti.com/lit/an/slva704/slva704.pdf.

[3] https://www.idt.com/products/power-management/wireless-power/introduction-to-wireless-battery-charging.

# 5
# Battery Protection Circuit Consideration

This chapter will discuss battery protection circuits in single and multicell batteries used in portable consumer electronic devices. While the high-level requirements of battery protection circuits used in larger format batteries are similar, the implementation of the protection circuits are typically different from protection circuit implementations in smaller form factor batteries used in portable consumer electronic devices.

## 5.1 Need for a Protection Circuit

Li-ion cells have a well-defined set of operating conditions. Using these cells outside these well-defined operating conditions increases the risk of a cell failure. This cell failure can be nonenergetic (benign) or energetic. Quite often, a Li-ion cell will fail in a benign manner. This may be exhibited as a cell that can no longer be charged or discharged, a cell that loses capacity in an accelerated manner, or a cell that is permanently disabled because of the operation of its internal protection component (e.g., the current interrupt device in a cylindrical cell). In some instances, a Li-ion cell may experience an energetic failure where the cell fails exothermically. An exothermic failure of the cell is referred to as the cell going into thermal runaway. Although infrequent, energetic failures of a Li-ion cell where the cell goes into thermal runway can be destructive and pose a fire hazard. As an example, Figure 5.1 shows a single-cell Li-ion battery that went into thermal runaway due to an external short-circuit condition.

Thermal runaway of a Li-ion cell occurs when the heat generated within the cell exceeds the heat dissipated by the cell. It is not uncommon for the surface temperature of a cell that goes into thermal runaway to exceed 600°C to 700°C during the thermal runaway event. As an example, Figure 5.2 shows the surface temperature of the single-cell Li-ion battery that experienced a thermal runaway due to an external short-circuit fault (Figure 5.1 shows the battery during the short-circuit condition).

The failure of a Li-ion cell, whether nonenergetic or energetic, can occur due to a variety of reasons. The probability of failure increases if the cell is operated outside its rated specifications. Figure 5.3 details some of the causes of a Li-ion cell's failure. At a high level, the causes of a Li-ion cell failure can be categorized into the following:

**Figure 5.1** Remnants of a Li-ion battery pack that went into thermal runaway due to an external short-circuit condition.

## 5.1 Need for a Protection Circuit

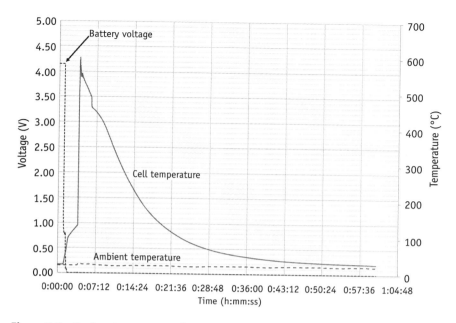

**Figure 5.2** Surface temperature of a 3-Ah Li-ion battery during an external short-circuit test.

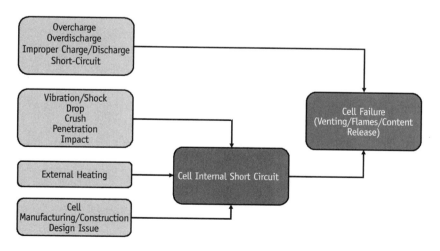

**Figure 5.3** Causes of Li-ion cell failure.

- *Electrical abuse*: cell overcharge, overdischarge, overcurrent charge/discharge, and short circuit;

- *Mechanical abuse:* excessive vibration, shock, drop, crush, penetration, and impact;
- *Thermal abuse:* elevated temperature storages and excessive external heating;
- Cell manufacturing/construction related issues.

Some of the factors that can cause a Li-ion cell's failure are as follows:

- *External short circuit*: A short-circuit condition in a fully charged multicell Li-ion battery can generate high peak currents. (As an example, a 2 Ah Li-ion cylindrical cell with a cobalt dioxide based positive electrode may generate peak short-circuit currents in excess of 50A. Cells designed for high-current applications can generate higher currents due to a comparatively lower internal impedance.) Under worst-case conditions, this can lead to cell venting with the release of flammable electrolyte, generation of toxic gases, or even a rupture of the cell enclosure.
- *Cell manufacturing defects:* The cell assembly process requires precision, control, and repeatability. Defects may be introduced within the cell during the assembly process, which can eventually lead to a cell internal short circuit condition. Cell defects introduced during the manufacturing/assembly process are most often related to the relative positioning of critical cell components. Some examples include
    - Improper cell tab positioning (folding and routing of tabs, tab overhang etc.);
    - Improper cell tab insulation;
    - Winding misalignments resulting in positive/negative electrode registry problems;
    - Crushing of the cell's electrodes.

A review of the cell's construction and frequently an audit of the cell's manufacturing plant are typically performed to identify whether the manufacturing process or cell construction has elements which increases the risk of a cell fault.

## 5.1 Need for a Protection Circuit

- *Cell charging algorithm:* Charging a Li-ion cell is a precise operation requiring features that control when and how the cell is charged. Li-ion cells are usually charged using the constant current-constant voltage (CC-CV) charge profile, which involves charging the cell at a constant current until its voltage reaches the predetermined limit (typically 4.1 or 4.2V) followed by a constant voltage charge state until the current decreases to a predetermined low value (see Chapter 4 for additional details). No trickle charging[1] is performed on Li-ion cells. Once fully charged, trickle charging Li-ion cells can result in the oxidization of the electrolytic solvents due to the high potential, physical and chemical degradation of the positive electrode material, and plating of metallic lithium at the negative electrode, which can eventually lead to a cell internal short-circuit condition.
- *Cell overcharge*: Overcharging a Li-ion cell increases the likelihood of the cell becoming thermally unstable.
- *Cell overdischarge*: Overdischarging a Li-ion cell can result in the oxidation of the copper current collector on the negative electrode, which leads to copper dissolution into the electrolyte. As the overdischarged cell is recharged, the dissolved copper redeposits in regions of the cell capable of reducing it back to copper metal. This can reduce cell performance by, for example, blocking access to active electrode material or clogging the pores of the separator membrane. If the Li-ion cell is overdischarged frequently, dendrite growth may start to occur between the negative and positive electrodes, which can eventually lead to an internal short within the cell.
- *Charging outside rated temperature:* Charging a Li-ion cell outside its rated temperature increases the risk of a cell failure during operation.

As seen in Figure 5.3, operation of a Li-ion cell outside its rated parameters for charge, discharge, and temperature, increases the

---

1. Trickle charging refers to a condition where a charge voltage is continuously applied to a cell. This is different from a condition where a cell is cycled (charged and discharged) continuously.

risk of the cell's failure in the field. For this reason, Li-ion cells, when used in applications, must contain dedicated protection circuits that will protect the cells from foreseeable electrical/thermal/mechanical abuse conditions and also prevent the operation of the cells in the application outside its rated specifications. Not only must protection circuits be designed to protect the Li-ion cells in an application, the application must also ensure that the charge and discharge infrastructure for the Li-ion cell does not stress the cell or increase the probability of a cell failure. The design of the battery protection circuit or the charge/discharge infrastructure in the application depends on a variety of factors, including the number of cells used in the application, the capacity of the cells, and the requirements of the application.

The battery protection circuit itself is typically a part of the Li-ion battery and is incorporated in an enclosure with the cells. The requirements and ultimate design of this protection circuit depends on a number of factors and varies considerably from one application to another.

## 5.2 Single-Cell Battery Packs

A large variety of applications ranging from smartphones, toys, and digital cameras typically utilize a Li-ion battery that contains a single Li-on cell. The battery protection circuit in single-cell Li-ion batteries is typically attached directly to the cell terminals. This circuit is often called a PCM. Single-cell PCMs typically contain a single protection integrated circuit (IC). This protection IC monitors the Li-ion cell's voltage and the charge and discharge current, and provides protection against cell overcharge, overdischarge, overcurrent, and external short-circuit conditions.

The PCM in single-cell batteries is often designed with two metal-oxide-semiconductor field-effect transistors (MOSFETs) that are controlled by the protection IC to terminate the charge and discharge current.[2] These MOSFETs are called the charge FET (C-FET) and the discharge FET (D-FET). The protection IC controls these MOSFETs as follows:

---

2. The protection IC may also permanently disable the battery pack if the cell voltage drops below a preset threshold to prevent an overdischarged cell from being used in the device.

- The C-FET is turned off when
  - The cell's voltage rises above a preset threshold;
  - The charge current exceeds above a preset threshold.
- The D-FET is turned off when
  - The cell's voltage drops below a preset threshold;
  - An elevated discharge current (exceeding a predesigned threshold) is detected;
  - A battery external short-circuit condition is detected.
  - The protection IC in single-cell battery packs does not always use a dedicated current sense resistor to monitor the charge and discharge current. The IC in some applications relies on the voltage drop across the C-FET and D-FET to determine the charge/discharge current. This allows for the protection IC to be used for battery packs with cells that have a range of charge/discharge current capabilities. The on-resistance of the C-FET and D-FET can be selected to set the overcurrent thresholds for charge and discharge.

Figure 5.4 shows an example of a simple PCM design in a single-cell Li-ion battery. The circuit in Figure 5.4 contains the following:

- The protection IC that monitors the cell voltage and the charge/discharge current and controls the C-FET and the D-FET.
- RC filtering circuit (R1 and C) connected between the cell and the power supply terminal of the protection IC. The filtering circuit provides a stable cell voltage reading to the protection IC. The same protection IC terminal can be used to provide power to the IC for operation. Resistor R1 also prevents the Li-ion cell from being exposed to a short circuit in the event of a failure on the PCM (for example a short-circuit failure of capacitor C).
- The protection IC monitors the voltage drop between the ground and the pin connected to resistor R2, and uses this information to determine the magnitude of the charge/discharge current to/from the cell. This information is

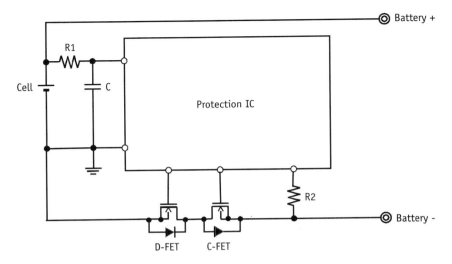

**Figure 5.4** Example connection scheme for a single-cell battery protection IC.

used to terminate the currents if they exceed a predesigned threshold.

In addition to the components shown in Figure 5.4, single-cell Li-ion PCMs may also include the following:

- Filter capacitors across the C-FET and D-FET. PCM designs often incorporate two series connected capacitors for this filtering. The use of a single capacitor for filtering increases the risk of a complete bypass of the protection circuit if the capacitor fails as a short-circuit.
- A negative temperature coefficient (NTC) thermistor that monitors the temperature in the vicinity of the cell and communicates this information to a charge circuit, which is typically located in the device that the battery powers. This allows the charger circuit to determine the temperature of the cell and ensure that the cell is only charged within its rated temperature.
  - NTC thermistors exhibit a decrease in electrical resistance with increasing temperature. The resistance value of a NTC thermistor is typically referenced at 25°C. For NTC thermistors used in single-cell batteries, this value abbre-

viated as 'R25' typically lies between 100 Ω and 100 kΩ. The other important NTC parameter for consideration when selecting a NTC for use in a single-cell battery PCM is the NTC's resistance versus temperature characteristics (R/T curve). The R/T curve is typically a nonlinear, negative exponential function [1]. However, interpolation equations are typically used to describe the R/T curve. The Steinhart-Hart equation is an empirical expression that has been determined to be the best mathematical expression for resistance temperature relationship of NTC thermistors. The most common equation is [2]

$$T = \frac{1}{A + B\ln(R) + C\left[\ln(R)\right]^3}$$

where $T$ is in degrees Kelvin and $A$, $B$, and $C$ are coefficients that are derived by calibrating at three temperature points and then solving three simultaneous equations. The uncertainty associated with the Steinhart-Hart equation is very small in the temperature range from 0°C to 260°C (this temperature range is suitable for Li-ion cells). This allows the use of a lookup table when using NTC thermistors to determine the thermistor's resistance at different temperatures. This makes NTC thermistors particularly suitable for Li-ion battery applications.

- Infrastructure that a charging circuit uses to identify the connected battery. This infrastructure can be as simple as a fixed resistor or can include sophisticated circuits with ICs that identify the battery by manufacturer, chemistry, or other identifying parameters. The purpose of this infrastructure is to allow a charger to identify a battery before charging. This reduces the risk of the use of imitation battery packs in an application.
- Single-cell Li-ion batteries also often include a secondary overcurrent/overtemperature component that is connected between one of the cell terminals and the PCM. This component is often either a PTC device or a thermal fuse. The purpose of this component is to provide secondary overcur-

rent/overtemperature protection to the cell and to protect the cell against a fault on the PCM itself. Although both the PTC and thermal fuse provide overcurrent and overtemperature protection, the PTC is resettable while the thermal fuse permanently disables the battery pack once its trip condition is reached. Regardless of whether a PTC or a thermal fuse is used in an application, the trip conditions are chosen so as not to conflict with the protection provided by the PCM. For example, if the PCM turns the D-FET off when the discharge current exceeds 10A for 10 ms, the PTC or thermal fuse will be chosen such that it does not trip at these current levels but at higher currents, or for the same current magnitudes but for longer current durations.

If the PTC or thermal fuse is chosen so that it does not overlap with the protection settings of the PCM, the triggering of either the PTC or the thermal fuse is an indicator of a failure on the PCM. For this reason some designers prefer the use of a thermal fuse rather than the resettable PTC to permanently disable the battery pack once the PCM is no longer fully functional.

Figure 5.5 shows a high-level block diagram showing the components of a typical single-cell Li-ion battery. These components are often incorporated in a single enclosure.

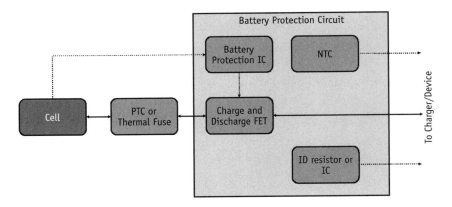

**Figure 5.5** Typical single-cell Li-ion battery pack.

## 5.3 Multicell Battery Packs

A large number of portable consumer electronic devices utilizing Li-ion batteries contain multiple cells (laptops, tablets, etc.). The cells can be connected both in series and parallel in these applications. A "yPxS" designation is often used to describe the battery pack's design. As an example, a 2P3S Li-ion battery contains six cells with two cells connected in parallel and three such pairs of cells connected in series. Similarly a 1P4S battery contains four cells connected in series. The use of both series and parallel connected cells in the battery increases the requirements for the PCM. The PCM in multicell batteries must monitor the voltage of each individual parallel connected cell string to provide both overcharge and overdischarge protection at the cell level. Similar to single-cell Li-ion batteries, the PCM in multicell batteries is typically attached directly to the cells via bus bars (or wires). In addition, the voltage of each individual parallel cell combination is communicated to the PCM through voltage sense wires (the voltage sense wires are typically separate from the bus bars or wires that carry the charge and discharge current. In addition to communicating the cell voltage information to the PCM, voltage sense wires are also used for cell balancing in applications with multiple cells connected in series).

Multicell battery PCMs typically contain the following:

- Two levels of individual parallel cell combination overcharge protection.
    - The two levels of overcharge protection are provided by two independent ICs on the protection circuit;
    - The second level of overcharge is typically designed to permanently disable the battery.
- Individual parallel cell combination undervoltage protection.
- Two levels of overcurrent protection for both charge and discharge (one of the two levels of protection is typically permanent (e.g., a fuse) and disables the battery preventing any further charge and/or discharge).

- Circuits to communicate with the charger circuit typically incorporated in the application that the battery powers.
    - This communication may include battery identification, the charge current magnitude, the maximum and minimum voltage of the cell pairs, the charge voltage, and current that the battery requires from the charger;
    - Communications may also include battery state of charge information, age of battery (i.e., the number of charge and discharge cycles that the battery has been exposed to), and battery date of manufacturer.
- Often the PCM also has the ability to store historical information, such as the maximum and minimum recorded temperature, parallel combination cell voltage, and charge and discharge current over the life of the battery. This information can be helpful when performing root cause analysis of a battery that has failed in the field.
- Cell balancing circuits to improve battery life and performance by maintaining the voltage balance across all series connected cells. Even if cells are closely matched when a battery is assembled, cell imbalance tends to increase over the life of the battery. Reasons for this cell imbalance growth can include the different operating temperatures of the cells due to temperature gradients in the battery and a variation of aging processes within the individual cells in the battery. The two commonly used techniques for cell balancing are
    - Passive balancing where cells that have reached a high voltage in comparison to their neighbors are discharged by switching in resistors in parallel to the cells to dissipate some of the storage energy;
    - Active balancing where charge is shuffled from one cell to another to equalize the charge on the cells using active elements (capacitors or inductors).

Figure 5.6[3] shows an example of a high-level block diagram of a multicell PCM (the figure shows a 2P3S battery configuration). A PCM design of this type is common in batteries used in tablets

---

3. This figure does not show the cell balancing circuit.

## 5.3 Multicell Battery Packs

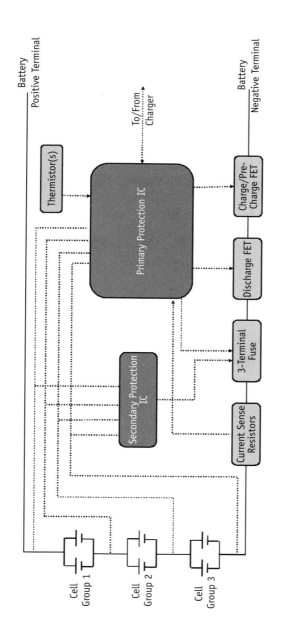

**Figure 5.6** High-level block diagram of multicell battery protection circuit.

and laptops. The PCM contains two independent protection ICs. The primary protection IC performs the bulk of the functions on the PCM. This IC monitors the cell temperatures (multicell battery packs utilize multiple thermistors to monitor the temperature at different locations in the battery), charge/discharge currents, the cell pair voltages, the health of the charge and discharge FETs and so forth. It also communicates with the charger circuit.

The secondary protection IC often only provides secondary overcharge protection functionality. It monitors the individual cell pair voltages and blows the three-terminal fuse if the voltage of any cell pair in the battery increases above the overcharge threshold. Often, the primary protection IC is also capable of blowing the three-terminal fuse under some fault conditions. For example, if the primary protection IC detects a failure of the charge or discharge FET (this would be detected because the primary protection IC would measure current flowing to/from the cells even though it has turned the charge or discharge FET off), it will blow the 3-terminal fuse to terminate the charge or discharge current.

The 3-terminal fuse in the battery pack is connected to the protection ICs via a FET (Figure 5.7). The primary and secondary protection ICs control a FET that is connected between one of the fuse terminals and the opposite end of the cell stack to which the fuse is connected. When either the primary or the secondary protection IC decide to turn this FET on, current flows through the heater of the three-terminal fuse causing a short circuit of the entire cell stack. This current flow melts and blows the three-terminal fuse and permanently disables the battery.

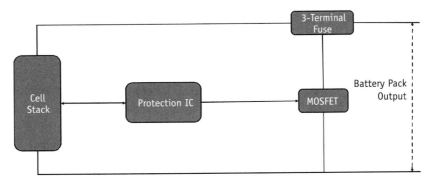

**Figure 5.7** Example connection scheme to the three-terminal fuse.

## 5.4 Large Battery Packs

Larger Li-ion battery packs, such as those used in telecommunication systems or electric vehicles, have more sophisticated battery protection circuits that perform a number of additional functions typically not required in smaller battery packs. These battery packs may contain the following features:

- A modular design with multiple protection ICs that communicate with each other and monitor the voltage of each individual parallel cell group in the battery pack;
- Contactors instead of charge/discharge FETs to handle the relatively larger operating currents;
- Cell balancing circuits that balance the voltages of the series connected cell groups during operation;
- Manual safety disconnects that de-energize the battery pack outputs to prevent electric shock hazards;
- Thermal management systems to ensure equal distribution of cell temperatures in the battery pack;
- Circuits to determine/track the state of health of the battery pack;
- Reinforced enclosures to provide mechanical integrity and protection to the cells.

Larger battery packs with nominal voltages of several hundred volts introduce another factor that must be addressed: electric shock hazards. Several UL and IEC standards list dc voltage magnitudes that are considered to be hazardous live. These UL standards are predominantly geared toward the safety of electrical equipment and hence address shock hazards associated with exposed energized parts of electrical equipment. For example, UL 61010-1 [3] defines hazardous live as follows:

> Hazardous live: capable of rendering an electric shock or electric burn in normal condition or single fault condition.

The standard states the following about hazardous live voltage levels (emphasis added):

> *Values above the following levels in normal condition are deemed to be hazardous live.*
> *The a.c. voltage levels are 30V r.m.s. and 42.4V peak and the dc level is 60V. For equipment intended to be used in WET LOCATIONS, the a.c. voltage levels are 16V r.m.s and 22.6V peak or 35V d.c.*

UL 60950-1 [4] on the other hand states the following (emphasis added):

> Steady state voltages up to 42.4V peak, or 60V d.c. are not generally considered as hazardous under dry conditions for an area of contact equivalent to a human hand.

Given the above information, systems utilizing Li-ion battery packs with nominal voltages that exceed 35V d.c. must give consideration to issues related to electric shock hazards. This may necessitate additional protection components on the PCM to ensure that a failure on the PCM does not expose a user to the risk of an electric shock.[4] A detailed review and discussion of the PCM requirements for large battery packs is outside the scope of this book.

## 5.5 Battery and PCM Specifications

The PCM specifications should be carefully evaluated once a PCM design has been selected to ensure that the PCM provides adequate protection to prevent the cell in the battery from operating outside its rated specifications under all foreseeable normal operating and fault conditions. At a minimum, the parameters listed in Table 5.1 should be obtained from the cell and battery specifications or other documentation supplied by the battery manufacturer. These specifications should then be matched with the specifications of the PCM to ensure that the PCM is designed to provide adequate

---

4. Appendix 5A provides further information on electrical shock hazards associated with DC voltages.

**Table 5.1**
Battery Specifications[1]

| 1 | Cell Specifications | Single-Cell Battery Packs | Multicell Battery Packs |
|---|---|---|---|
| 1.1 | Nominal capacity (minimum) | 1100 mAh (0.2C discharge) | 2200 mAh (0.2C discharge) (Individual cell capacity × number of cells in parallel) |
| 1.2 | Typical capacity | 1150 mAh (0.2C discharge) | 2300 mAh (0.2C discharge) |
| 1.3 | Nominal voltage | 3.7V | 11.1V (3.7V × number of series cell combinations) |
| 1.4 | End voltage | 3.0V | 9.0V (3.0V × number of series cell combinations) |
| 1.5 | Charging method | CC–CV (constant current–constant voltage) | CC–CV (constant current–constant voltage) |
| 1.6 | Standard charge rate | Constant current of 0.5C | Constant current of 0.5C |
|  |  | Constant voltage of 4.2 ± 0.05V | Constant voltage of 12.6 ± 0.15V (4.2 ± 0.05V × number of series cell combinations) |
|  |  | 0.05 C rate cutoff current (stop charge) | 0.05 C rate cutoff current (stop charge) |
| 1.7 | Fast charge rate | Constant current: 1C rate | Constant current: 1C rate |
|  |  | Constant voltage: 4.2 ± 0.05V | Constant voltage: 12.6 ± 0.15V (4.2 ± 0.05V × number of series cell combinations) |
|  |  | 0.05C rate cutoff current (stop charge) | 0.05 C rate cutoff current (stop charge) |
| 1.8 | Discharge current | 1C rate maximum continuous | 1C rate maximum continuous |
| 2 | **Temperature** |  |  |
| 2.1 | Charge | 0°C to 45°C | 0°C to 45°C |
| 2.2 | Discharge | −20°C to 60°C | −20°C to 60°C |
| 2.3 | Storage[2] | −20°C to 45°C | −20°C to 45°C |
| 3 | **Appearance** |  |  |
| 3.1 | Markings | Manufacturer name |  |
|  |  | Part number, model number, and equivalent designation |  |
|  |  | Electrical ratings of voltage and capacity |  |
|  |  | Battery type |  |
|  |  | Date of manufacture |  |
|  |  | User warnings |  |
| 4 | Shipping requirements |  |  |
| 4.1 | Charge on battery when shipped | 40% to 60% of rated capacity |  |

1. This table provides the typical specifications for a 1.1 Ah lithium cobalt dioxide based Li-ion cell, and for a 6-cell (2 parallel 3 series (2P3S) configuration) battery, which are used as reference. All typical values for the protection settings listed in this table are for a 1.1 Ah cell. These values will change for cells with different chemistries, capacities and ratings.

2. Storage temperature requirements vary depending upon the length of storage time. The cell specifications should be consulted to determine the storage temperature requirements for the application. Prolonged storage may require the cells to be periodically recharged to prevent cell overdischarge.

protection to the Li-ion cell(s) and meets the battery specifications provided by the manufacturer.

Table 5.2 provides typical protection settings and features of PCMs for single and multicell Li-ion battery packs. This table assumes that the cell and battery specifications are as listed in Table 5.1. Note that protection levels may be set slightly higher than battery specifications. For example, in Table 5.1, constant voltage charging is to be accomplished with a constant voltage of 4.2 ± 0.05V, indicating a maximum voltage of 4.25V for the charging circuit. However, in Table 5.2 the overcharge detection voltage is set to 4.25 ± 0.03V. Thus, protection could be activated across a range

Table 5.2
PCM Specifications

| | Specifications | Single-Cell Battery Packs | | MultiCell Battery Packs |
|---|---|---|---|---|
| 1 | Operating Temperature | −20°C to 60°C, 95% Humidity | | |
| 2 | Overcharge voltage protection | Detection voltage | 4.25V ± 0.03V | Voltage of highest parallel cell combination > 4.25V ± 0.03V |
| | | Release voltage | 4.07V ± 0.05V | Voltage of highest parallel cell combination < 4.07V ± 0.05V |
| 3 | Overdischarge voltage protection | Detection voltage | 2.7 ± 0.1V | Voltage of lowest parallel cell combination < 2.7 ± 0.1V |
| | | Release voltage | 3.0V ± 0.1V | Voltage of lowest parallel cell combination > 3.0 ± 0.1V |
| 4 | Over current protection | Detection current | > 2A | Detection current > 4A |
| 5 | Short circuit protection | Detection current | > 7.5A | Detection current > 15A |
| | | Delay time | < 30 ms. | < 30 ms. |
| | | Release voltage | Load release | Load release |
| 6 | Secondary protection | PTC thermistor/Thermal fuse/Fuse | | Independent secondary protection IC for overcharge protection and three-terminal fuse |
| 7 | Allowable charge temperature | Typically managed by the charger | | 0°C to 45°C |
| 8 | Allowable discharge temperature | — | | −20°C to 60°C |

of values from 4.22 to 4.28V. Such a discrepancy is not unusual, and is related to limitations of component tolerances. When such discrepancies are identified, the cell manufacturer is typically consulted to determine if the suggested PCM specifications are acceptable.

Table 5.1 and Table 5.2 provide examples of some of the battery and PCM specifications that must be reviewed and finalized for a battery system. The tables are not meant to be an exhaustive list of all specifications for single and multicell batteries and their PCMs in portable consumer electronic applications. Examples of other items that should be considered include

- Ensuring that the PCM provides protection that is adequate for the chemistry of the Li-ion cell used in the application. For example, the nominal voltage of Li-ion cells with an iron phosphate positive electrode (3.2V) is different from the nominal voltage of a Li-ion cell with a cobalt dioxide based positive electrode (3.7V). The different nominal and fully charged voltages for these cells necessitate a different threshold for overcharge protection, overdischarge protection, and so forth.
- The number of temperature sensors that are used in the PCM/charger circuitry and the location of these sensors to ensure the best possible monitoring of cell temperatures in the battery.
- The minimum cell voltage at which the PCM goes into sleep mode and the PCM operating current in the sleep mode. The lack of a sleep mode or elevated operating currents in the sleep mode increase the likelihood of cell overdischarge during extended periods of storage.
- The need and design of cell balancing infrastructure in a PCM to balance cells when multiple cells are connected in series.
- The infrastructure/protocol used for communication between the PCM and the charger circuit (typically located in the device that is powered by the battery).

## 5.6 PCM Design Review Tools

Several tools are available to battery pack designers for reviewing PCM designs. These reviews should be performed early during the design/development process so that appropriate changes to the design can be made before too much investment into the production of the PCMs has occurred. Some of the tools available include design failure modes and effects analysis (DFMEAs), process hazard analysis (PHA), and hazard and operability (HAZOP) analysis.

A DFMEA is frequently the tool of choice when evaluating the design of a battery pack PCM. A DFMEA is a structured approach that ensures that potential PCM failure modes and their associated causes have been considered and addressed in the product design. The DFMEA is a methodical approach used for identifying risks introduced in a new or even a modified PCM design. The DFMEA initially identifies PCM design functions, failure modes and the effects of these failure modes on the overall safety and reliability of the battery pack with a corresponding severity ranking of the effect. Then, causes and the mechanisms of the failure mode are identified. High probability causes, indicated by the occurrence ranking, may drive action to prevent or reduce the cause's impact on the failure mode. The detection ranking highlights the ability of specific tests to confirm that the failure mode/causes are eliminated. The DFMEA also tracks improvements through risk priority number (RPN) reductions. By comparing the before and after RPN, a history of improvement and risk mitigation can be tracked.

Performing a DFMEA

- Allows for an early identification of the ways that the PCM can fail;
- Prioritizes potential PCM failures so that corrective/preventative actions and/or a redesign of the PCM can be accomplished;
- Reduces warranty costs that would occur due to PCM failures in the field;

- Enables for the eventual release of a more reliable product that will result in higher end-user satisfaction.

One of the most important elements of performing a useful DFMEA is to ensure the formation of a proper team for the process. DFMEA team members will often include design engineers, manufacturing and process engineers, quality, reliability, and field services engineers, and so forth. In addition, often members of the QA/QC, packaging and marketing teams also get involved in the DFMEA process.

## 5.7 PCM[5] Construction and Assembly Issues

A visual inspection of the PCM/BMU and its integration with the cell(s) should be performed to ensure that the layout is such that it does not increase the probability of a fault. The layout of any traces should be designed in conformance with the guidance in industry standards, such as UL 60950-1 and IPC 2221. The layout and assembly requirements for multicell battery packs requires a review of several additional features not commonly needed for single-cell battery packs.

### 5.7.1 Single-Cell Battery Packs

For single-cell battery packs, the following items require attention:

- *Battery connector:* Typically, the battery connector will consist of multiple pins. These pins will include connections for the positive and negative terminals of the battery, pins for communication with the charger, and so forth. The connector design should be such that the positive and negative terminals are at two opposite ends of the connector (i.e., are as far apart as possible on the connector).

---

5. The term PCM will be used to refer to the protection circuit for single-cell battery packs while the term BMU will be used to refer to the protection circuit for multicell battery packs in this section of the chapter.

- *PCM connection:* For Li-ion polymer cells, the PCM and tabs connecting the cell to the PCM should be insulated from the cell pouch. This is typically accomplished through insulation of the cell tabs where they exit the cell and insulation of the PCM with layers of paper and/or insulating tape.
- *Battery pack wires*: Any battery pack wires that provide a connection between the battery pack and the device that it powers must be routed away from and adequately insulated from the PCM and the cell tabs.
- *Temperature monitoring:* PCM designs that incorporate a thermistor to monitor the cell temperature should have the thermistor positioned in a manner that ensures accurate tracking of the cell temperature.
- *Cell tab separation:* The cell tab connections on the PCM should be separated as much as possible.

### 5.7.2 Multicell Battery Packs

The various cells in multicell battery packs are typically connected together via bus bars that allow current to flow from the cells to the BMU. In addition, voltage sense wires communicate the voltages of each parallel cell combination to the BMU. The voltage sense wires are designed to only carry small monitoring and cell balancing currents. For multicell battery packs, the following items require attention:

- *Temperature monitoring:* Multicell battery packs typically utilize one or more temperature sensors to monitor the ambient temperature around the cells. The temperature distribution within the battery depends on a number of factors, such as heat generated within the cells, the heat dissipation mechanism in the battery, the placement of cells within the battery, and so forth. Thermal characterization techniques are often used for larger format batteries to determine the most suitable location for the thermal sensors in the battery. The locations of the thermal sensors should be such that they maxi-

mize the ability of the sensors to monitor overtemperature anywhere in the battery.
- *Battery connector:* Typically, the battery connector will consist of multiple pins. These pins will include connections for the positive and negative terminals of the battery, and pins for communications with the charger. The connector design should be such that the positive and negative terminals are at the two opposite ends of the connector (i.e., are as far apart as possible on the connector).
- Cell connections to the BMU
  - Voltage sense wires connected to each parallel cell group provide cell voltage information to the BMU. The routing of these wires is especially important and should be such that the wires do not move inside the battery enclosure and are routed such that the possibility of an electrical short circuit inside the battery is minimized. In addition, adequate separation should be provided between these wires at locations where the wires are soldered to the BMU circuit board to prevent the occurrence of resistive faults.
  - Current limiting resistors are typically utilized on the BMU to limit the current that may flow through the voltage sense wires in the event of a fault on the BMU. The location of these current limiting resistors on the BMU should be reviewed to ensure that they are as close as possible to the location where the voltage sense wires connect to the BMU.
  - If the BMU utilizes filter capacitors for filtering the cell voltages, these filter capacitors may act as single points of failure and result in a short circuit of multiple cells in the battery if the capacitors fail short-circuit. The location and type of capacitors used for filtering requires attention in the BMU design.
- *Bus bars:* Bus bars are typically used to connect the cells together and to connect the cell stack to the BMU.
  - The routing of the bus bars should be inspected to ensure that the bus bars do not have the potential to damage any cell insulation during operation. In addition, multiple lay-

ers of insulation are typically used between the cells and bus bars when the cell is at a different potential from the bus bar.
- The solder joint/connection of the bus bars to the BMU should be inspected to ensure that there are no sharp edges, which can damage insulation during operation
- The spot welding of the bus bars to the cell tabs/terminals should be inspected to determine the quality of the spot welds. Excess heat application during the spot welding process can damage the cell and increases its probability of failure in the field.
- *Solder quality:* The quality of the solder connections (specifically in and around the battery output connector) should be inspected for cold solder joints, cracks, and/or contamination.
- *Flux residue/contaminants:* The BMU should be inspected for excessive flux residue or signs of external contaminants. Excessive flux residue or contamination can result in short-circuit paths between components.

## 5.8 Cell Failure Predictions

Some experimental PCM/BMU designs are also incorporating cell failure prediction systems to predict a cell's failure and permanently disable the battery pack when the prediction indicates imminent cell failure. As was discussed earlier, there are several reasons that can cause a Li-ion cell to fail. Several techniques have been proposed in the literature to predict a Li-ion cell's failure in the field. These techniques have involved the following:

- The use of sensors wrapped around individual cells to detect changes in cell dimensions as an indication of an imminent failure;
- Monitoring the charge profile and using a change in the charge profile as an indication of the possibility of a cell failure;

- Changes to the charge profile proposed in the literature have included an extension in the charge current taper time and a change in the charge efficiency;
  - A noisy voltage and/or current signal in the charge profile as an indication of a microshort in the cell, which can eventually cause a cell failure.
- Monitoring the open-circuit voltage of a cell during rest and using an accelerated drop in the open-circuit voltage of the cell (i.e., accelerated self-discharge rate) as an indication of a faulty cell that can eventually fail;
- Monitoring the cell temperature and disabling the cell/battery pack if the cell temperature rises above a preset threshold;
- Monitoring the cell capacity and using an accelerated capacity degradation as an indication of a faulty cell that may eventually fail;
- Monitoring the cell impedance and using a rise in the cell impedance above a preset threshold as an indication of a faulty cell that may eventually fail.

### 5.8.1 Cell Resistance as a Predictor of Failure

A review of the available literature indicates that several researchers have attempted to use a change in the resistance of a cell as a predictor of a cell's eventual failure. A literature search identified two different approaches to predict cell failure

- Rate of resistance increase
  - Research has indicated that the internal resistance of a Li-ion cell increases slowly before a cell goes into thermal runaway. Once a cell goes into thermal runaway a much higher rate of rise in cell resistance is observed. As an example, in a study of a large format NCM type Li-ion cell an ARC test was performed to characterize the internal resistance of a cell before the cell went into thermal run-

away. The test performed indicated that the internal resistance of the tested cell increased gradually from 20- to 60- m$\Omega$ before thermal runaway. The internal resistance of the cell increased to 370 m$\Omega$ when the cell went into thermal runaway during the test [5].
- This indicates that it may be possible to monitor the cell resistance and use a rapid increase in cell resistance as an indicator of a cell's failure
- However, it is not known if these cell resistance characteristics are the same for the different Li-ion cell types (e.g., NCA type cells, NMC type cells, and LiCoO$_2$ type cells). This requires further research.
- The authors are unaware of any applications where this technique is used for predicting a cell's imminent failure in an application or of the practical considerations that may impact the success of this method.
- Single-point impedance based diagnostic
  - One researcher has employed a single-point impedance based diagnostic to monitor the state-of-health of individual Li-ion cells in a multicell battery pack as a way of detecting when a cell is overcharged.
  - This research has determined that the internal impedance of the tested cell at a particular frequency is independent of the cell SOC over its normal operating window. However, this internal impedance changes if the cell is overcharged.
  - The proposed detection strategy relies on generating a baseline of cell impedance both during normal operating conditions and under overcharge conditions. A change in the cell impedance to levels that are associated with an overcharge condition is then used to make a determination that the cell has been overcharged so that appropriate action can be taken to prevent a cell failure.
  - This change in impedance is used as a means of detecting the overcharged condition of a cell.

- It is not known if the cell impedance characteristics discussed in the study will vary for different Li-ion cell types (e.g., NCA type cells, NMC type cells, and $LiCoO_2$ type cells).
- The authors are unaware of any applications where this technique is used for predicting a cell's imminent failure in an application or of the practical considerations that may impact the success of this method.

## 5.9 Testing PCMs[6]

Once the design and assembly of a PCM is completed, it is important to perform testing on battery samples to ensure that the PCM operates as intended and provides the desired protection under all foreseeable fault conditions. Different single-cell and multicell battery pack configurations demand different protocols to be followed when evaluating the PCMs. While the type of tests performed will depend on a number of factors, such as the PCM design, the size of the battery, and the application that the battery is used in, the following category of electrical tests are often performed when evaluating PCMs regardless of the size of the battery in which the PCM is incorporated.

- *Charge cut-off test:* Overcharging a Li-ion cell can cause the cell to go into thermal runaway. The failure of a cell when overcharged is typically more energetic than when the cell goes into thermal runaway in a fully charged condition because the cell has comparatively more energy when it is overcharged. The aim of this test is to determine the ability of the PCM to prevent cell overcharge under all foreseeable operating and fault conditions. This test simulates a failure of the charger circuit that results in the charger outputting an elevated charge voltage to the battery pack.
- *Discharge cut-off test:* Overdischarging a Li-ion cell increases the probability of the cell failing during subsequent charge

---

6. The term PCM is used to refer to the protection circuit for both single and multicell batteries in this section.

cycles. The aim of this test is to determine the ability of the PCM to prevent cell overdischarge under all foreseeable operating and fault conditions. This test determines the cell voltage threshold at which the PCM terminates the discharge current. Often the system that is powered by the battery pack is designed to shut down and terminate the discharge current before the overdischarge voltage threshold is reached. This test simulates a failure of the voltage detection circuit in the device that is powered by the battery pack.

- *External short-circuit test:* An external battery short circuit can result in elevated current flow from the cells, which can cause damage to components on the circuit board, insulation damage, and cell overheating. The aim of this test is to determine the ability of the PCM to detect an external short-circuit current and turn the battery off to prevent elevated current flow. This test simulates a failure on the device that the battery pack powers that results in a short circuit of the battery pack.

- *Overcurrent charge:* Li-ion cells can only be charged at rates as specified by the cell manufacturer. This rate is typically specified as a percentage of the cell's capacity. Elevated charging currents increase the probability of cell failure. The aim of this test is to determine the ability of the PCM to prevent the cells from being charged at elevated currents under all foreseeable operating and fault conditions. This test simulates a failure of the charger circuit that results in the charger outputting an elevated charge current to the battery pack.

- *Overcurrent discharge:* Similar to charge currents, Li-ion cells have a maximum discharge current rating. Exceeding this discharge current rating increases the probability of a cell failure. The aim of this test is to determine the ability of the PCM to prevent the cells from being discharged at elevated currents (especially at currents in excess of the cell's maximum discharge current rating) under all foreseeable operating and fault conditions. This test simulates a failure on the device that the battery pack powers, which results in an overload condition at the output of the battery pack.

- *External short-circuit test (primary protection failure):* This test simulates a double fault condition: a failure of the device that the battery pack powers, resulting in a short-circuit of the battery pack and a failure of the primary overcurrent protection circuit in the PCM. The aim of the test is to ensure that the secondary protection built into the PCM provides adequate protection to the cells.

- *Cell overcharge test (primary protection failure):* This test also simulates a double fault condition: a failure of the charger circuit that results in the charger outputting an elevated charge voltage to the battery pack and a failure of the primary overcurrent protection circuit in the PCM. The aim of the test is to ensure that the secondary protection built into the PCM provides adequate overcharge protection to the cells (this test is often only necessary for PCMs in multicell batteries as single-cell batteries do not always have an active secondary overcharge protection circuit).

- *Cell imbalance test:* This test is only required for batteries with multiple cells connected in series. The rate at which Li-ion cells age in an application can vary from one cell to another. This results in the formation of an imbalance condition in multicell battery packs due to the different cells aging at different rates over time. As a result of this imbalance, cells that age more tend to charge and discharge at elevated rates compared to other cells in the battery pack resulting in dissimilar states of charge among the cells. The PCM should ensure that this natural and expected cell imbalance does not result in an overcharge or overdischarge of any cell in the battery pack. The aim of this test is to ensure that the PCM can prevent a cell from operating outside its rated voltage under the worst case imbalance condition.

- *Voltage sense line failure test:* For multicell battery packs, voltage sense lines are used to communicate the voltage of each cell pair in the battery pack to the PCM. The PCM should be designed to permanently disable the battery pack if it loses voltage information from any parallel cell group in the battery pack. The aim of this test is to determine the response of the PCM to a loss of voltage information from any cell group in the battery pack.

The list of tests presented above is not meant to be exhaustive. The exact tests that are performed will depend on the application, the size of the battery, and the design of the PCM. The above list only provides a guide of the types of test that must be considered when determining the suitability of a PCM's design for an application.

## 5.10 Summary

The Li-ion cell has a well-defined range of operating conditions for voltage, current, and temperature. Failures generally occur when a cell is operated outside its specifications or exposed to mechanical, electrical, or thermal abuse conditions. A cell failure could lead to a cell that

- Cannot be charged and/or discharged;
- Leaks electrolyte;
- Vents and releases gases;
- Overheats;
- Experiences thermal runaway.

Any application utilizing Li-ion cells must be designed with a PCM that protects the cell under all foreseeable fault conditions. The designs of these PCMs are inherently application dependent. It is important to also exhaustively review and test the PCMs once a design has been finalized to ensure that the PCM works as intended in the application.

## Appendix 5A: Electric Shock Hazards

### Introduction

"For a given current path through the human body, the danger to persons depends mainly on the magnitude and duration of current flow."[6] However, when designing measures of protection against electrical shock, it is not practical to use time/current values to determine the level of protection needed. The criteria typically used for evaluating the risk of electrical shock is touch voltage. Touch voltage is defined as the "product of the current through the body called touch current and the body impedance" [6]. The relationship between the current and voltage is not typically linear because the impedance of the human body varies with touch voltage and also the different parts of the body that the current flows through. In addition, the value of the body impedance depends on factors such as duration of current flow, magnitude of touch voltage, frequency, degree of moisture on the skin, surface area of contact, pressure exerted, and temperature.

According to IEC TS 60479-1, "accidents with direct current are much less frequent than would be expected from the number of d.c. applications, and fatal electrical accidents occur only under very unfavorable conditions, for examples, in mines. This is partly due to the fact that with direct current, the let-go of parts gripped is less difficult and that for shock durations longer than the period of the cardiac cycle, the threshold of ventricular fibrillation is considerably higher than for alternating current."

### Thresholds

When discussing electrical shock hazards, standards typically divide the hazards into the following thresholds:

- Threshold of perception;
    - This is the minimum value of current that causes any sensation for the person through which it is flowing.
- Threshold of reaction;
    - This is the minimum value of current which causes involuntary muscular contraction.
- Threshold of let-go;
    - This is the maximum value of touch current at which a person holding electrodes can let go of the electrodes.
- Threshold of ventricular fibrillation.
    - This is the minimum value of touch current through the body that causes ventricular fibrillation.

### Touch Voltage Thresholds for Ventricular Fibrillation

Table 5.3 [7] provides the perception/reaction, muscular reactions and ventricular fibrillation threshold current values for different current paths through the body. As can be seen in Table 5.3, the threshold current magnitudes for ventricular fibrillation depend strongly on the path that the current takes through the body.

The magnitude of current flow through the body is also a function of the body's contact area with the energized part as well as whether the conditions are dry or wet. The following conditions are typically defined in standards when addressing electric shock hazards:

- *Dry:* condition of the skin of a surface area of contact with regard to humidity of a living person at rest under normal indoor environmental conditions;
- *Water-wet:* condition of the skin of a surface area of contact being exposed for 1 minute to water of public water supplies;
- *Saltwater-wet:* condition of the skin of a surface area of contact being exposed for 1 minute to a 3% solution of salt water.

Table 5.4 [7] provides the DC touch voltage thresholds for ventricular fibrillation for different contact scenarios if contact is maintained for a long duration.[7] According to Table 5.4, the lowest

**Table 5.3**
Current Threshold Values for the Three Types of Thresholds for Current Flow for 10 Seconds

| Type of Threshold | Current Path | Current Magnitude (mA) |
|---|---|---|
| Perception/reaction | Hand-to-hand | 2 |
| | Both hands to feet | 2 |
| | One hand to feet | 2 |
| Strong muscular reactions | Hand-to-hand | 25 |
| | Both hands to feet | 25 |
| | One hand to feet | 25 |
| Ventricular fibrillation | Hand-to-hand | 350 |
| | Feet-to-both-hands | 140 |
| | Seat-to-one-hand | 200 |

---

7. IEC/TR 60479-5 defines long duration as approximately 10 seconds.

**Table 5.4**
DC Touch Voltage Threshold for Ventricular Fibrillation for Direct Current
(Contact Maintained for a Long Duration)

|  | Saltwater-wet | | | Water-wet | | | Dry | | |
| --- | --- | --- | --- | --- | --- | --- | --- | --- | --- |
|  | Large Contact[1] | Medium Contact[2] | Small Contact[3] | Large Contact | Medium Contact | Small Contact | Large Contact | Medium Contact | Small Contact |
| Hand-to-hand (350 mA) | 263V | 351V | 467V | 264V | 353V | 470V | 264V | 264V | 470V |
| Both-hands-to-feet (140 mA) | 68V | 121V | 220V | 75V | 143V | 223V | 87V | 143V | 223V |
| Hand-to-seat (200 mA) | 83V | 126V | 201V | 85V | 127V | 203V | 85V | 127V | 203V |

1. Large contact area: defined as a surface area of 82 cm$^2$.
2. Medium contact area: defined as a surface area of 12.5 cm$^2$.
3. Small contact area: defined as a surface area of 1 cm$^2$.

DC touch voltage threshold value for ventricular fibrillation of 68V occurs under saltwater wet conditions for a large contact area when the current flows from both hands to feet.

## References

[1] https://www.sensorsmag.com/embedded/negative-temperature-coefficient-thermistors-part-i-characteristics-materials-and.

[2] https://www.ametherm.com/thermistor/ntc-thermistors-steinhart-and-hart-equation.

[3] UL 61010-1: Electrical Equipment for Measurement, Control, and Laboratory Use–Part 1: General Requirements, 2012, p. 50.

[4] UL 60950-1: Information Technology Equipment–Safety–Part 1: General Requirements, 2007, p. 17.

[5] Feng, X., et al., "Thermal Runaway Features of Large Format Prismatic Lithium Ion Battery Using Extended Volume Accelerating Rate Calorimetry," *Journal of Power Sources,* Vol. 255, 2014, pp.294–301.

[6] IEC TS 60479-1: Effects of Current on Human Beings and Livestock–Part 1: General Aspects, 2005-07, pp. 21, 93.

[7] IEC/TR 60479-5 Technical Report: Effects of Current on Human Beings and Livestock–Part 5: Touch Voltage Threshold Values for Physiological Effects, 2007-11, pp. 11,13, and 19.

# 6

# Industry Standards and Testing

A large number of industry standards have been devised to characterize the response of Li-ion cells, battery packs, and devices incorporating Li-ion battery systems to different abuse conditions. These standards are typically developed through a consensus process, which relies on participation by representatives from regulatory bodies, manufacturers, industry groups, consumer advocacy organizations, insurance companies, and other key safety stakeholders. Often the technical committees developing requirements for product safety standards rely less on prescriptive requirements and more on performance tests simulating reasonable situations [1].

## 6.1 Commonly Used Standards to Evaluate Li-Ion Cells and Batteries in Portable Consumer Electronic Devices

Several Li-ion battery related standards have been developed to evaluate the safety of Li-ion batteries under different abuse conditions. These standards are typically specified based on the application and range from standards for household and commercial batteries, abuse standards for electric vehicles, requirements for batteries used in e-cigarette devices, and so forth. This chapter will focus on Li-ion battery related standards that have been developed for portable consumer electronic devices. Table 6.1 and 6.2 provide a high-level overview of some of the Li-ion cell and battery related certification requirements and standards commonly used around the world (the tables are not meant to be an exhaustive list of all relevant Li-ion battery system standards around the world, but are meant to provide the reader with an overview of the types of requirements for the batteries around the world).

Manufacturers of portable consumer electronic devices typically rely on the original battery manufacturer to obtain the relevant certifications for the battery systems. The market or region where the product will be sold and used primarily dictates which certifications are needed or will be acquired. The standard relevant for the device that incorporates the Li-ion battery will also often dictate which standards to use for evaluating battery within the product. As an example, IEC 60601 (a series of technical standards for medical electrical equipment) requires that if a medical device utilizes a rechargeable Li-ion battery, the battery needs to be tested in accordance with IEC 62133[1] [2]. The standards also typically specify the minimum number of cells to be tested. Although testing to standards and proving compliance is a major step in the product life cycle, assessing the risks and quality of the product should be done throughout this life cycle, right from the inception to end of life. Hence, some manufacturers perform periodic spot checking (testing of certain number of cells) as a way to ensure continued compliance.

### 6.1.1 UN Transportation Requirements

In several countries, Li-ion cells or batteries cannot be transported either individually or installed in a device until they pass the tests listed in the United Nations (UN) Recommendations on the

---

1. Secondary cells and batteries containing alkaline or other nonacid electrolytes – safety requirements for portable sealed secondary lithium cells, and for batteries made from them, for use in portable applications.

**Table 6.1**
Li-ion Cell Related Standards and Certifications

|  | US | International | Europe | China | Korea |
|---|---|---|---|---|---|
| Certifications/ Marks | UL | IEC; IECEE | CE | CQC | KC |
| Voluntary/ Mandatory | Voluntary | Voluntary | Voluntary | Mandatory | Mandatory |
| Factory Inspection | Yes | Not required | Yes | Yes | Not required |
| Certification Validity | No expiration assuming no change in product | Contingent on standard upgrade | 10 Years | As long as routine factory inspection is passed | No expiration assuming no change in product |
| Standards Applied | UL 1642 | IEC 62133 | EN 62133 | GB 31241-2014 | Part 2, annex 05 of self-regulatory confirmation |

Transportation of Dangerous Goods Manual of Tests and Criteria. A summary of UN Testing Requirements from the *Sixth Revised Edition of the Manual of Tests and Criteria* (Effective 2015) is provided in Table 6.3.

### 6.1.2 Underwriters Laboratories Testing Requirements

Underwriters laboratories (UL) has published a number of standards related to Li-ion batteries. UL 1642 (Standard for Lithium Batteries) and UL 2054 (Standard for Household and Commercial Batteries) are two of the common standards used to evaluate Li-ion batteries used in portable consumer electronic products in the United States.

#### 6.1.2.1 UL 1642: Standard for Lithium Batteries

A summary of some of the UL 1642 requirements is provided in Table 6.4. This standard requires tests to be performed on both new cells and cells that have been aged via charge/discharge cycling. Most of the tests are performed on five samples.

#### 6.1.2.2 UL 2054: Standard for Household and Commercial Batteries

A summary of some of the UL 2054 requirements is provided in Table 6.5. UL 2054 is used to test battery packs while UL 1642 is used

Table 6.2
Li-ion battery Pack Related Standards and Certifications

| | US | Canada | Germany | Japan | Russia | China | Korea |
|---|---|---|---|---|---|---|---|
| Certifications/Mark | UL | ULC | UL (DE), GS | DENAN | GOST | CQC | KC |
| Voluntary/Mandatory | Voluntary | Voluntary | Voluntary | Mandatory | Mandatory | Mandatory | Mandatory |
| Factory Inspection | Yes | Yes | Yes | Not required | Yes | Yes | Not required |
| Standards Applied | IEC 60950-1 with UL 2054 | CSA 60950-1 with UL 2054 | EN 60950 and EN 62133 | DENAN Ordinance, Article 1, Appendix 9 | GOST 62133 | GB 31241-2014 | Part 2, annex 05 of self-regulatory confirmation |

## 6.1 Commonly Used Standards to Evaluate Li-ion Cells and Batteries

**Table 6.3**
UN Transportation T-Tests

| Clause | Test | Protocol | Criteria |
|---|---|---|---|
| UN 38.3.4.1 | Test T.1 – Altitude Simulation | Batteries stored at a pressure of 11.6 kPa or less for at least 6 hours at ambient temperature (20± 5°C) | No leakage No venting, disassembly, rupture or fire OCV drop < 10% |
| UN 38.3.4.2 | Test T.2 – Thermal Test | Rapid thermal cycling between high (72 ± 2°C) and low (–40± 2°C) storage temperatures | No leakage No venting, disassembly, rupture or fire OCV drop < 10% |
| UN 38.3.4.3 | Test T.3 – Vibration | Vibration exposure: sinusoidal waveform with a logarithmic sweep from 7- to 200- Hz and back to 7 Hz in 15 minutes; 12 cycles, 3 perpendicular mounting positions | No leakage No venting, disassembly, rupture or fire OCV drop < 10% |
| UN 38.3.4.4 | Test T.4 – Shock | Shock exposure: half-sine shock, 150g peak acceleration, 6ms pulse duration, 3 shocks in positive and negative directions in each of three perpendicular mounting positions (total of 18 shocks) | No leakage No venting, disassembly, rupture, or fire OCV drop < 10% |
| UN 38.3.4.5 | Test T.5 – External Short Circuit | Short circuit of less than 0.1 ohm at 55 ± 2°C | No disassembly, rupture or fire within 6 hours. of test. Cell temperature does not exceed 170°C |
| UN 38.3.4.6.2 | Test T.6 – Impact | 15.8 ± 0.1 mm diameter bar, at least 6 cm long, type 316 stainless steel bar, placed across cell center, and a 9.1 ± 0.1 kg mass is dropped onto the bar from 61 ± 2.5 cm height | No disassembly or fire within 6 hrs. of test. Cell temperature does not exceed 170°C |
| UN 38.3.4.6.3 | Test T.6 – Crush | Cell crushed between two flat surfaces. Crushing done at a speed of 1.5 cm/s until applied force reaches 13 ± 0.78 kN, voltage drops by at least 100 mV or cell is deformed by more than 50% from its original thickness | No disassembly or fire within 6 hrs. of test. Cell temperature does not exceed 170°C |
| UN 38.3.4.7 | Test T.7 – Overcharge | Over current (2× manufacturer's recommended maximum) and overvoltage charge (For 18V batteries or less, charge to the lesser of 22V or 2× recommended charge voltage. For >18V batteries, charge to 1.2× recommended charge voltage) | No disassembly or fire within 7 days of test |
| UN 38.3.4.8 | Test T.8 – Forced Discharge | Each cell force discharged by connecting it in series with a 12V DC power supply at an internal current equal to maximum allowable discharge current | No disassembly or fire within 7 days of test |

**Table 6.4**
UL 1642 Tests

| Clause | Test | Protocol | Criteria |
|---|---|---|---|
| 10 | Short-circuit test | Short circuit battery through maximum resistance of $80 \pm 20 m\Omega$ at both $20°C \pm 5°C$ and $55°C \pm 5°C$. | No explosion/fire |
| 11 | Abnormal charging test | Overcurrent charging test (constant voltage, current limited to 3× specified max charging current); testing at $20°C \pm 5°C$; testing of fresh and cycled (conditioned) batteries; 7 hours duration | No fire/explosion |
| 12 | Forced-discharge test | Fully discharged cell is force-discharged by connecting it in series with fully charged cells of the same kind. Once connected, resultant battery pack is short-circuited through maximum resistance of $80 \pm 20\ m\Omega$ | No fire/explosion |
| 13 | Crush test | Battery is crushed between two flat plates to an applied force of $13 \pm 1$ kN ($3,000 \pm 24$ lb.); | No fire or explosion |
| 14 | Impact test | $15.8 \pm 0.1$ mm diameter bar is placed across a battery; a $9.1 \pm 0.46$ kg ($20 \pm 1$ lb.) weight is dropped on to the bar from a height of $24 \pm 1$ inches ($61 \pm 2.5$ cm) | No fire or explosion |
| 15 | Shock test | Three shocks applied with minimum average acceleration of 75g; peak acceleration between 125 and 175g; shocks applied to each perpendicular axis of symmetry; testing at $20°C \pm 5°C$ | No fire or explosion No venting or leakage |
| 16 | Vibration test | Simple harmonic vibration with amplitude of 0.8 mm applied to batteries in three perpendicular directions; frequency is varied between 10 and 55 Hz at the rate of 1 Hz/min | No fire or explosion No venting or leakage |
| 17 | Heating test | Battery placed into an oven initially at $20°C \pm 5°C$; oven temperature is raised at a rate of $5°C \pm 2°C$/minute to a temperature of $130°C \pm 2°C$; the oven is held at $130°C \pm 2°C$ for 10 minutes, then the battery is returned to room temperature; | No fire or explosion |
| 18 | Temperature cycling test | Battery is cycled between high and low temperatures: 4 hours at $70 \pm 3°C$, 2 hours at $20 \pm 3°C$, 4 hours at $-40 \pm 3°C$, return to $20 \pm 3°C$ and repeat the cycle a further nine times; testing of fresh and cycled cells | No fire or explosion No venting or electrolyte leakage |
| 19 | Low pressure (altitude simulation) test | Batteries stored at an absolute pressure of 11.6 kPa (1.68 psi) for 6 hours at $20°C \pm 3°C$ | No fire or explosion No venting or electrolyte leakage |
| 20 | Projectile test[1] | Cell heated until it explodes or has ignited and burned out | No part of exploding cell or battery shall penetrate the wire screen |

1. Applicable to user replaceable batteries only.

## Table 6.5
## UL 2054 Tests

| Clause | Test | Protocol | Criteria |
|---|---|---|---|
| 10 | Abnormal charging test | Overcurrent charging test (constant voltage, current limited to 3× specified max charging current); testing at 20°C ± 5°C; testing of fresh and cycled (conditioned) batteries; 7 hours duration | No explosion/fire No chemical leaks caused by cracking, rupturing, or bursting of battery casing |
| 11 | Abusive overcharge test | Overcurrent charging test (constant voltage, current limited to 10× $C_5$ charging current); testing at 20°C ± 5°C; testing of fresh and cycled (conditioned) batteries; test run until battery temperature reaches steady state conditions or returns to room ambient | No fire or explosion |
| 12 | Forced discharge test | Fully discharged cell is force-discharged by connecting it in series with fully charged cells of the same kind. Once connected, resultant battery pack is short circuited through a resistance load of 80 ± 20 mΩ | No fire/explosion |
| 13 | Limited power source test | Characterize battery response to varying output load conditions | Batteries that meet the requirements are eligible to include the marking LPS to indicate that they are considered limited power sources |
| 14 | Crush test | Battery is crushed between two flat surfaces to an applied force of 13 ± 1.0 kN (3,000 ± 24 lb.); | No fire or explosion |
| 15 | Impact test | 15.8 mm diameter bar is placed across the center of the sample; a 9.1 ± 0.46 kg (20 lb.) weight is dropped on to the sample from a height of 24 ± 1 inches (61 cm) | No fire or explosion |
| 16 | Shock test | Three shocks applied with minimum average acceleration of 75g; peak acceleration between 125 and 175g; shocks applied to each perpendicular axis of symmetry; testing at 20°C ± 5°C | No fire or explosion No venting or leakage |
| 17 | Vibration test | Simple harmonic vibration with amplitude of 0.8 mm applied to cells in three perpendicular directions; frequency is varied between 10 and 55 Hz at the rate of 1 Hz/min; testing at 20°C ± 5°C | No fire or explosion No venting or leakage |

**Table 6.5** (continued)

| Clause | Test | Protocol | Criteria |
|---|---|---|---|
| 20 | Mold stress relief test | 7-hour exposure to 70°C in a full-draft circulating-air oven | No evidence of mechanical damage that would result in damage to cells or protective devices<br>No fire or explosion<br>Battery enclosure should not crack, warp, or melt such that access to cell or protective devices is possible |
| 21 | Drop impact test | Drop each of the three samples (three times) from height of 1m onto concrete surface; testing at 20°C ± 5°C; | No fire or explosion<br>No venting or leakage 6 hours after the test<br>Battery enclosure should not crack such that access to cell or protective devices is possible |
| 22 | Projectile test | Battery heated until it explodes or has ignited and burned out | Should not ignite cheesecloth layer |
| 23 | Heating test | Battery placed into an oven initially at 20°C ± 5°C; oven temperature is raised at a rate of 5°C ± 2°C/minute to a temperature of 130°C ± 2°C; the oven is held at 130°C for 10 minutes, then the battery is returned to room temperature | No fire or explosion |
| 24 | Temperature cycling test | Battery is cycled between high and low temperatures: 4 hours at 70°C ± 3°C, 2 hours at 20°C ± 3°C, 4 hours at -40°C ± 3°C, return to 20°C ± 3°C and repeat the cycle a further nine times; testing of fresh and cycled cells | No fire or explosion<br>No venting or electrolyte leakage |

to test cells. UL 2054 requires that cells meet the requirements of UL 1642. As with UL 1642, most of the tests are required to be performed on five battery samples. Since these tests are performed on battery packs that typically have a protection circuit built in, most tests in this standard recommend repeating the test if a protection component of the battery pack operates to terminate the fault condition. In the event this happens, the standard recommends that the test be performed at a level where the protection circuit does not operate. As an example, if the overcurrent protection setting operates during the external short circuit test, the UL standard recommends that the test be repeated at an overload condition below the level at which the overcurrent protection in the battery pack will operate. This is done to ensure that the battery pack and its components can handle the worst-case discharge currents at which the overcurrent protection circuit will not operate.

Standard UL 62133 (Secondary Cells and Batteries Containing Alkaline or Other Non-Acid Electrolytes - Safety Requirements for Portable Sealed Secondary Cells, and for Batteries Made From Them, for Use in Portable Application) was released in 2015 with the intention of making the standard fully harmonized with IEC 62133.

### 6.1.3 IEC Standards

IEC standard CEI/IEC 61960 (Secondary cells and batteries containing alkaline or other non-acid electrolytes - Secondary lithium cells and batteries for portable applications) provides a description of standard cell designations. This standard also provides procedures for assessing cell performance under a variety of conditions, such as various temperatures, various discharge rates, and after extended cell cycling. The aim of the standard is to allow battery purchasers to compare performance of different cells under a single set of tests. This standard does not include any safety tests.

The IEC publishes a standard that specifically addresses safety requirements for rechargeable cells and batteries: IEC 62133 (Secondary cells and batteries containing alkaline or other non-acid electrolytes – Safety requirements for portable sealed secondary cells, and for batteries made from them, for use in portable applications – Part 2: Lithium systems). Many of the tests described in this standard are very similar to those described in UL standards. IEC 62133 also includes a set of design and manufacturing requirements. Some of these requirements and tests listed in this standard are summarized in Table 6.6.

IEC publishes another standard that specifically addresses safety requirements for rechargeable cells and batteries during transportation: IEC 62281 (Safety of primary and secondary lithium cells and batteries during transport) [3]. This standard also addresses safety during transportation for recycling or disposal.

## Table 6.6
IEC 62133 Requirements and Tests

| Clause | Requirement | Criteria |
| --- | --- | --- |
| 5.2 | Insulation and wiring | Wiring and insulation resistance between positive terminal and externally exposed area $\geq 5$ M$\Omega$ at 500 VDC, 60s after applying the voltage; Internal wiring and insulation are sufficient to withstand maximum operating conditions Creepage and clearance distances are adequate; Mechanical integrity of internal connections accommodate foreseeable misuse |
| 5.3 | Venting | Battery case construction to allow mechanism to relieve excessive internal pressure; Any encapsulation used does not cause battery to overheat during normal operation or inhibit pressure relief |
| 5.4 | Temperature/ voltage/ current management | Abnormal temperature rise should be prevented by design Batteries are designed to be within temperature, voltage, and current limits as per cell specifications; Batteries are provided with specifications and charging instructions for equipment manufacturers |
| 5.5 | Terminal contacts | Size and shape of terminals should be such that they can carry the maximum anticipated current; Conductive materials used on contact surfaces should have good mechanical strength and corrosion resistance; Terminals contacts are arranged to minimize risk of short circuits |
| 7.2.2 | Case stress at high ambient temperature | 7 hour exposure to 70°C in air circulating oven internal protection device and cells shall not be exposed |
| 7.3.2 | External short circuit | Short circuit of less than 0.1 $\Omega$ at 20°C and 55°C, 24-hour duration, no fire or explosion |
| 7.3.3 | Free fall | Dropped three times from height of 1m onto concrete or metal floor in random orientations, no fire or explosion |
| 7.3.4 | Thermal abuse | Fully charge cells soaked at 20°C $\pm$ 5 °C for 1 hour in an oven, temperature then raised at a rate of 5°C $\pm$ 2°C/min to 130°C $\pm$ 2 °C and kept at this temperature for 30 minutes; no fire or explosion |
| 7.3.8.1 | Vibration | Vibration exposure: sinusoidal motion with amplitude of 0.8 mm. Frequency varied from 7 to 200 Hz and back to 7 Hz in approximately 15 minutes. Three perpendicular mounting positions, testing repeated 12 times for a total of approximately 3 hours for each position; no fire, explosion, rupture, leakage, or venting |
| 7.3.8.2 | Mechanical shock | Three shocks applied with peak acceleration of 150g with a half sine pulse of 6 ms duration; shocks applied in each direction of each perpendicular axis of symmetry for a total of 18 shocks; testing at 20°C |
| 7.3.9 | Forced internal short circuit | A forced internal short circuit test for cylindrical and prismatic cells shall not cause a fire. The test is performed on the winding core removed from a charged cell. An L-shaped nickel particle is inserted between the positive coated area and the negative coated area. A temperature controlled chamber and special press equipment (pressing jig) is needed for the test. The pressing jig is moved down at a speed of 0.1 mm/s until an internal short circuit is detected (as a voltage drop). |

### 6.1.4 IEEE Standards

#### 6.1.4.1 Std. 1725: IEEE Standard for Rechargeable Batteries for Cellular Telephones[2]

In the United States, the IEEE standards are voluntary. However, cell phone carriers, through CTIA (The Wireless Association) have mandated compliance with IEEE Std. 1725 (IEEE Standard for Rechargeable Batteries for Cellular Telephones) to their suppliers. IEEE Std. 1725 emphasizes that battery safety is a function of a number of interrelated components: the cells, the battery, the host device (i.e., charger), the power supply accessories, the user, and the environment. IEEE Std. 1725 establishes that the "responsibility for total system reliability is shared between the designers/manufacturers/suppliers of the subsystems and the end user."

The standard requires a design analysis for a system using tools such as FMEA or fault tree analysis. The standard also requires that cells and batteries comply with UN and UL 1642 requirements, and also recommends testing to UL 2054 and IEC 62133. The standard includes some additional testing that goes beyond the standard tests already described in previous sections. In addition, the standard provides a guide on desirable features that should be taken into consideration when designing single-cell Li-ion batteries and chargers. Table 6.7 provides a list of desirable features for Li-ion battery chargers while Table 6.8 provides similar requirements for battery packs as identified by the standard.

### 6.1.5 Comparing Standards

The different Li-ion technology related test standards all contain a series of similar abuse tests for the cells/batteries. Table 6.9 [1] summarizes the abuse tests in the various standards often used to evaluate Li-ion batteries used in portable consumer electronic devices. While the exact test protocol used for each abuse condition may vary between the standards, the intent of the tests in the various standards is similar.

As an example, Table 6.10 summarizes the protocol for the vibration test as specified in the different standards.

---

2. This standard focuses on single-cell battery packs.

**Table 6.7**
IEEE Std. 1725 Requirements for Li-ion Chargers[1,2]

| Requirement | Specifications |
|---|---|
| Safety features | The charger shall not disable or degrade the safety features inside the battery pack, the device interface should prevent reverse polarity connection. |
| Electrostatic discharge | Specific electrostatic discharge (ESD) protection shall be included in the system design, testing, and qualification. |
| Mating of pins | The host device and battery connections shall mate properly and be capable of good electrical contact throughout their respective product lifetimes. |
| Pin separation | Power and ground pins shall be sufficiently separated to minimize the possibility of an accidental short circuit. |
| Current ratings | Conductors and connectors shall have the proper current rating for current load with adequate margin as determined by the system manufacturer/supplier. |
| Metallurgy consideration | The charger connector pin metallurgy shall be compatible with the host connector pin metallurgy to minimize corrosion and resistance changes over the life of the system. |
| Connector strength | The host connector to the battery shall be mechanically robust. |
| Mating force | Adequate mechanical force between the electrical contact points shall be maintained to minimize any fretting or other electrical degradation of the contact. |
| Shock and vibration effects | The host device shall withstand shock and vibration caused by normal usage and shall not propagate faults to the battery and cells when they are installed in the system. |
| Foreign objects | Precautions shall be taken to minimize the potential for foreign objects and/or liquids to enter the host device and cause a short circuit either during the manufacturing process or end-use operation. |

1. Some of the requirements listed in this table have been taken from IEEE Std. 1625: IEEE Standard for Rechargeable Batteries for Portable Computing. This standard addresses multicell series connected batteries. However, the requirements listed in this standard are also applicable to single-cell Li-ion energy storage systems.
2. The host device includes the battery charger circuit.

Figure 6.1 shows an example test fixture for a vibration test, which includes a shaker table and an upright fixture to mount and hold the test cells in $X$, $Y$, and $Z$ directions.

## 6.2 Other Tests

### 6.2.1 Forced Internal Short-Circuit Test

IEC 62133 specifies the forced internal short circuit test as a design evaluation test. The test is listed as a "country specific test, which is only applicable to France, Japan, Korea and Switzerland and is

**Table 6.8**

IEEE Std. 1725 Requirements for Li-ion Batteries[1]

| Requirement | Specifications |
|---|---|
| Leakage protection | The battery pack should be designed to mitigate hazards from contamination of electronic circuits by electrolyte if cells leak. |
| Circuit layout | The design of the battery pack shall provide adequate runner spacing, soldering pad area size, and distance between solder pads as well as separation between traces to minimize the occurrence of accidental short circuits. |
| Ambient thermal considerations | Thermal specifications of battery pack components shall not be exceeded when the pack is tested at the maximum charge and discharge currents with the pack ambient temperature elevated to the maximum temperature specification of the host. |
| Component specifications | Components, and materials used in the battery shall meet the minimum and maximum temperature requirements with adequate margin. |
| Thermal sensor design | The battery pack and/or host shall contain at least one thermal protection mechanism independent of any internal cell device or mechanism. |
| Electrical cell connections | Tabs, plates, and lead cables should be connected by welding and/or soldering to provide securer mechanical and electrical connections. Connections shall not be soldered directly to the cell. |
| Connection resilience | Appropriate spacing should be provided to prevent abrasion, wear, or damage to cable leads and/or connectors in the battery pack. Appropriate strain relief should be provided for cable leads and/or connectors in the battery pack. |
| Damage resilience | Electrical parts should have appropriate clearance to allow for situations where the battery pack is deformed by external mechanical force. |
| Cell vent | Battery pack construction shall not prevent cell vent gases from escaping |
| Circuit isolation | A circuit board in the battery pack, if used, should be mechanically isolated and secured to prevent unintended electrical connections. |
| Electrostatic discharge | Specific electrostatic discharge (ESD) tolerance shall be included in the battery pack design, testing, and qualification. |
| Welding | Welding shall only be applied in areas designated by cell manufacturer/supplier in accordance with agreed upon specifications. Critical processes such as welding shall have a quality control and maintenance plan to control the consistency of the assembly process and adherence to specifications. There shall be no damage to cell enclosure or cell case and critical cell design elements during welding and other operations. |
| Soldering process | Adequate means shall be provided to prevent solder balls, solder flashes, solder bridges, and foreign debris. IPC 2221A is one standard that may be applied. |

1. Some of the requirements listed in the table have been taken from IEEE Std. 1625: IEEE Standard for Rechargeable Batteries for Portable Computing. This standard addresses multicell series connected batteries. However, the requirements listed in this standard are also applicable to single-cell Li-ion energy storage systems.

**Table 6.9**
Common Abuse Tests in Different Standards

| Test Criteria/Standard | UL 1642 | UL 2054 | UN 38.3 | IEC 62133-2 | IEEE 1725 | JIS C 8714[1] |
|---|---|---|---|---|---|---|
| Abnormal Charge | ✓ | ✓ | ✓ | ✓ | ✓ | ✓ |
| Forced Discharge | ✓ | ✓ | ✓ | ✓ | ✓ | ✓ |
| External Short Circuit | ✓ | ✓ | ✓ | ✓ | ✓ | ✓ |
| Crush | ✓ | ✓ | — | ✓ | ✓ | ✓ |
| Impact | ✓ | ✓ | ✓ | — | ✓ | — |
| Shock | ✓ | ✓ | ✓ | ✓ | ✓ | ✓ |
| Vibration | ✓ | ✓ | ✓ | ✓ | ✓ | ✓ |
| Heating | ✓ | ✓ | — | ✓ | ✓ | ✓ |
| Temperature Cycling | ✓ | ✓ | ✓ | ✓ | ✓ | ✓ |
| Altitude Simulation | ✓ | — | ✓ | ✓ | ✓ | ✓ |
| Projectile | ✓ | ✓ | — | — | ✓ | ✓ |
| Drop | — | — | — | ✓ | — | — |
| Internal Short Circuit | — | — | — | ✓ | — | ✓ |

1. Safety tests for portable Lithium Ion secondary cells and batteries for use in portable electronic applications, Japanese Standards Association

**Table 6.10**
Vibration Test Protocols in Industry Standards

| Standard | UL 1642 | UL 2054 | UN 38.3/IEEE 1625 | IEC 62133-2 | IEEE 1725 |
|---|---|---|---|---|---|
| Level | Cell | Pack | Cell/Pack | Pack | Pack |
| Directions | Three axes | Three axes | Three axes | Three axes | Three axes |
| Vibration Mode | Simple harmonic | Simple harmonic | Sine, logarithmic | Sine, logarithmic | Simple harmonic |
| Frequency Range | 10–55 Hz | 1–55 Hz | 7–200 Hz | 7–200 Hz | 10–55 Hz |
| Maximum Displacement/Peak Acceleration | 1.6 mm | 1.6 mm | 1.6 mm/8g | 1.6 mm/8g | 1.6 mm |
| Duration | 90–100 minutes | 90–100 minutes | 12 cycles, 15 minutes each | 12 cycles, 15 minutes each | 90–100 minutes |
| Pass Requirements | No fire, vent, leak | No fire, vent, leak | No fire, vent, leak | No fire, explosion, rupture, vent, leak | No fire, explosion, rupture, vent, leak |

not required on lithium ion polymer cells" [4]. The forced internal short circuit test is also part of the Japanese standard JIS C 8714.

## 6.2 Other Tests

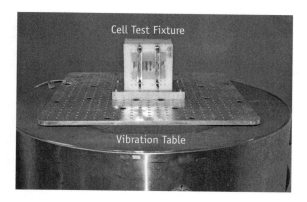

**Figure 6.1** Vibration test setup example.

The protocol for this test involves dismantling the cell and inserting a small sliver of metallic nickel between the cell's positive and negative electrodes. The cell is then reassembled and a pressing force is applied until a cell internal short is achieved. A cell passes this test if the forced internal short circuit condition does not cause a fire.

### 6.2.2 Nail Penetration Test

The nail penetration test is another test commonly used to simulate a cell internal short circuit condition. This test involves driving a metallic nail or rod through a charged Li-ion cell at a prescribed speed. This test is not part of the many Li-ion abuse standards typically used for testing Li-ion cells and batteries that are intended for portable consumer electronic devices. A methodology for this test is described in the U.S. Advanced Battery Consotium (USABC) test manual for electric and hybrid electric vehicle applications. This test is also listed in Section 4.3.3 of the SAE J2464 (Electric and Hybrid Electric Vehicle Rechargeable Energy Storage System (RESS) Safety and Abuse Testing) standard. The protocol for this test based on the USABC manual includes penetrating the cell or battery pack (or module) with a mild steel (conductive) pointed rod, electrically insulated from the test article, and driven at a rate of 8 cm/sec. The diameter of the rod is to be 3 mm when testing cells and 20 mm when testing modules and full battery packs.[3] As the parameters for the nail penetration test are not standardized, various different methodologies are used to conduct this test on cells intended for use in portable consumer electronic devices. As an example, Figure 6.2 shows a nail penetration test setup in which the nail penetration is performed at an elevated ambient temperature. In this test setup, the cell is heated using a heating jacket to increase the cell temperature to the desired test temperature before penetrating the nail. Testing performed has indicated that the cell response to penetration by a nail depends upon a number of factors including the nail material and diameter, the speed and depth of penetration, and the shape and taper angle of the tip of the metallic nail or rod.

### 6.2.3 Thermal Stability Test

A protocol for the thermal stability test is detailed in both the USABC standard and the SAE J2464 standard, both of which are geared toward testing large-format cells, modules, and battery packs. The basic idea of this test is to increase the temperature of the cell, module, or battery at a constant rate until a failure is achieved, which often is the thermal runaway of the test cell.

**Figure 6.2** Nail penetration test of a heated cell using a hydraulic drive.

---

3. FreedomCAR Electrical Energy Storage System Abuse Test Manual for Electric and Hybrid Electric Vehicle Applications, USABC, 2005.

### 6.2.4 Dent/Pinch Test

The nail penetration test is one way of trying to achieve a cell internal short circuit and characterizing the response of the cell to this short-circuit condition. Typically, the nail penetration test results in the short circuit of several electrode layers in the cell even if only a partial penetration test is performed. In addition, the cell enclosure is compromised during the test, which results in the escape of any gases from the compromised enclosure. Furthermore, the metal rod or nail that is used for the penetration may also act as a heat sink once it penetrates the cell. For this reason, other methodologies have been developed to try and create a short circuit within a cell and evaluate the response of the cell to this short-circuit condition. One such test is the dent/pinch test. One proposed test setup uses two spheres to apply concentrated coaxial loading on both sides of a prismatic cells [6]. A servohydraulic mechanical testing machine is used to control the applied pressure to the cells during the test. In this way the applied pressure can be released as soon as an internal short circuit within the cell develops. Other variants of this test setup include a dent test where the loading is applied using a blunt end of a metallic rod [7]. Once again, the use of a servohydraulic mechanical testing machine allows for the applied loading to be removed once an internal short circuit within the cell develops.

## 6.3 Predicting Li-ion Cell and Battery Shelf Life[4]

The ability to predict the life of a cell in an application is part of performing a complete functional evaluation of the Li-ion cell. When selecting a Li-ion battery for an application that requires the battery to operate reliably for a period of time, it is important to estimate the battery's performance given the operating conditions in the application. Selecting the wrong battery in these applications can result in expensive replacements as the batteries start to age prematurely and fail to meet the requirements of the application.

---

4. From [8].

Performing real life aging studies on a battery is an expensive and time-consuming task that may also be infeasible for some applications. Hence, over the years, many techniques and models have been proposed to estimate a Li-ion battery's life in an application. A very popular technique for battery life prediction uses extrapolation of experimental data gathered by performing short-term accelerated aging experiments to predict future performance of the battery in an application. The popularity of this technique is due to the simple accelerated aging experiments that are needed for predicting the battery's life in the application. Many different methodologies of accelerated aging and extrapolation of the accelerated aging data have been proposed in the literature. One such methodology relies on the Arrhenius equation, which assumes that the capacity degradation of Li-ion cells during storage is predominantly temperature-dependent. The methodology relies on predicting battery life based on accelerated aging tests that are typically performed for 12 to 16 weeks at different ambient temperatures. Battery life and capacity degradation are often listed in the battery specification. Battery manufacturers sometimes use similar life prediction methods to come up with this specification.

### 6.3.1 Capacity Degradation

For Li-ion batteries with graphitic negative electrodes, one of the dominant degradation mechanisms involves the growth of the solid-electrolyte interphase (SEI) layer that increases the cell's impedance and reduces its capacity as it consumes cyclable lithium from the cell [10]. The ambient temperature at which a cell operates and/or is stored can have a large impact on the cell's capacity degradation. A high ambient temperature during storage coupled with a high SOC accelerates the rate of capacity degradation in a cell. When considering capacity degradation due to cycling over the entire lifetime of the battery, losses due to storage should also be taken into account. The Arrhenius equation can be used to determine the impact of ambient temperature on calendar life.

### 6.3.2 Calendar Aging

To determine the aging characteristics of a 3-Ah Li-ion cell with a cobalt dioxide based positive electrode, calendar aging tests were performed. As part of these tests, the cells were stored at different temperatures and states of charge. Each test condition was performed with multiple cells to demonstrate the reproducibility of the experiment. The results showed a similar aging for cells tested under the same conditions. As an example, the mean capacity loss at 60°C for cells stored at different states of charge observed during the testing is as shown in Figure 6.3. The data indicates that the cells experience an increasing capacity loss with a higher SOC and temperature (this was discussed in detail in Chapter 1).

### 6.3.3 Fit Method

The Arrhenius equation can be used to obtain a lifetime model using aging tests. The Arrhenius equation describes the relationship

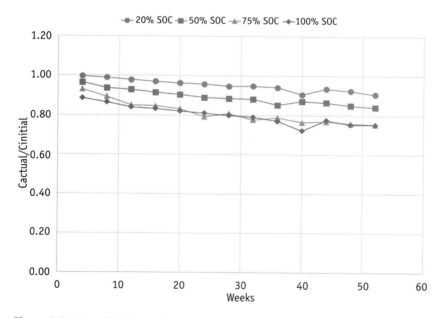

**Figure 6.3** Normalized capacity over 52 weeks at 60°C at different states of charge.

between the rate at which a reaction proceeds and its temperature. This equation is given as follows:

$$C_f = A * e^{\frac{-Ea}{RT}}$$

where

$C_f$ is the rate of change of some measured condition;

$A$ is a dimensionless factor related to the probability that change in the measured condition will occur;

$E_a$ is the activation energy required to change the measured condition;

$R$ is the ideal gas constant;

$T$ is temperature in degrees K.

Taking the natural logarithm of the Arrhenius equation yields:

$$\ln C_f = \left(\frac{-E_a}{R}\right) * \left(\frac{1}{T}\right) + \ln A$$

The capacity degradation at different temperatures (for a particular SOC) can be used to determine the activation energy ($E_a$) and the constant $A$ by plotting the natural logarithm of the capacity degradation as a function of inverse temperature.

### 6.3.4 Time Dependence

For a Li-ion cell with a carbon-based anode, such as the one that was used for the experiments, the literature indicates that the dominant calendar aging effect is the formation of the SEI. The SEI is built up of decomposition products of the electrolyte, consuming lithium during formation and increasing resistance through the growing layer thickness. The literature indicates that although there are different theories on SEI formation, they all lead to a square root of time dependency [9].

### 6.3.5 Summary

The Arrhenius equation provides a relatively simple method of predicting the shelf life of a Li-ion cell in the field. Simple capacity degradation measurements after storing the cells at different states of charge and ambient temperatures can be used with simple charge/discharge cycling experiments to predict the aging characteristics of a Li-ion cell/battery in an application in the field.

## Appendix 6A: International Organizations and Standards

### United Nations (UN)

The UN issues recommendations for the transport of dangerous goods worldwide along with the United States Department of Transportation (DOT), which defines shipping regulations for the U.S. Safety tests are defined in UN Test Manual (38.3) – *Manual of Tests & Criteria for Transport of Dangerous Goods (Lithium Metal & Lithium Ion Batteries)*.

### Underwriters Laboratories

UL is an independent product safety certification organization that, in conjunction with other organizations and industry experts, publishes consensus-based safety standards. For lithium batteries, some of the relevant standards are

*UL 1642*: Standard for primary (nonrechargeable) and secondary (rechargeable) lithium batteries for use as power sources in products,

*UL 2054*: Standard for household and commercial batteries,

*UL 2580*: Batteries for Use in electric vehicles,

*UL 2271*: Standard for Safety - Batteries for Use in light electric vehicle (LEV) applications [cells should comply with UL 1642].

### The International Electrotechnical Commission (IEC)

The IEC prepares publishes a number of standards that are used in a variety of industries. Battery-related standards published by IEC include the following:

*IEC 62133-2*: Standard for Secondary Cells for Portable Applications

*IEC 60086-4*: Standard for Safety of Primary Lithium Batteries

*IEC 62281*: Standard for Safety of Primary and Secondary Lithium Cells and Batteries During Transport

*IEC 61960*: Standard for Secondary Cells and Batteries containing Alkaline or other Non-Acid Electrolytes – Secondary Lithium Cells and Batteries for Portable Applications

*IEC 62660-3*: Standard for Secondary Lithium-Ion Cells for the Propulsion of Electric Road Vehicles

### The Institute of Electrical and Electronics Engineers (IEEE)

IEEE is an international nonprofit organization. The key Li-ion battery related standards published by IEEE include IEEE 1725 (Rechargeable Batteries for Cellular Telephones) and IEEE 1625 (Rechargeable Batteries for Multi-Cell Mobile Computing Devices).

### SAE International

*SAE J2929*: Electric and Hybrid Vehicle Propulsion Battery System Safety Standard - Lithium-based Rechargeable Cells

*SAE J2464*: Recommended Practice for Electric Vehicle Battery Abuse Testing

*SAE J2380*: Standard for Vibration Testing of Electric Vehicle Batteries

### Japanese Standards Association (JIS)

*JIS C 8714*: Safety Tests for Portable Lithium-Ion Secondary Cells and Batteries for Use in Portable Electronic Applications

### Other Standards

*ISO 12405-2*: Electrically Propelled Road Vehicles - Test Specification for Lithium-Ion Traction Battery Packs and Systems

*USABC:* Electric Vehicle Battery Test Procedures Manual

*ECE R100.02*: Regulation for Uniform Provisions Concerning the Approval of Vehicles with Regard to Specific Requirements for the Electric Power Train [Part II: Requirements of a Rechargeable Energy Storage System (REESS) with Regard to its Safety]

*FreedomCAR*: Electrical Energy Storage System Abuse Test Manual for Electric and Hybrid Electric Vehicle Applications

*BATSO 01:* (Proposed) Manual for Evaluation of Energy Systems for Light Electric Vehicle (LEV) — Secondary Lithium Batteries

## Appendix 6B: Common Tests in Industry Standards

### Shock Test

The shock test is a common test listed in a number of standards. Table 6.11 summarizes the test protocol for the shock test detailed in some of the standards commonly used to test Li-ion cells and batteries for use in portable consumer electronic devices.

At a high level the protocol for the shock test in the various standards requires that the cell or battery being tested is secured to the testing machine by means of a rigid mount that supports all mounting surfaces. Each test sample is then subjected to the shock condition along each of the three mutually perpendicular directions unless the test sample has only two axes of symmetry, in which case only two directions are tested. Each shock is typically applied in a direction normal to the face of the cell and the

Table 6.11
Shock Test Protocols in Industry Standards

| Standard | UL 1642 | UL 2054 | UN 38.3/ IEEE 1625 | IEC 62133-2 | IEEE 1725 |
|---|---|---|---|---|---|
| Level | Cell | Pack | Cell/Pack | Pack | Pack |
| Directions | Three axes, three shocks | Three axes, three shocks | Three axes, three shocks in + and − directions | Three axes, three shocks in + and − directions | Three axes, three shocks |
| Peak Acceleration | 125–175g | 125–175g | 150g | 150g | 125–175g |
| Duration | > 3 ms | > 3 ms | 6 ms | 6 ms | > 3 ms |
| Pass Requirements | No fire, vent, leak | No fire, vent, leak | No fire, vent, leak | No fire, explosion, rupture, vent, leak | No fire, explosion, rupture, vent, leak |

**Figure 6.4** Shock test setup example.

standards require that the cells do not vent, leak or catch fire during the test. Some standards also require that the open-circuit voltage of cells do not decrease below a certain amount during the test. Figure 6.4 shows an example test fixture for a shock test that includes a shock table and an upright fixture to mount and hold the test cells in the $X$, $Y$, and $Z$ directions. Custom cell mount fixtures are typically designed based on the cell size and format.

### Impact Test

The aim of the impact test is to evaluate the consequences of mechanical abuse (impact) on a cell or battery. As part of the impact test, the test standards require that the test sample be subjected to a single impact with a fixed weight and from a fixed height. The acceptable sample response in this test is for the test sample to not explode of catch fire. Figure 6.5 shows an example of an impact test setup. In this setup, the cell is placed on a flat surface and impacted by a weight.

### Crush Test

As implied by the name of the test, the aim of the crush test is to evaluate the response of a cell or battery to a crushing force.

*Appendix 6B: Common Tests in Industry Standards* 159

Figure 6.5  Impact test setup example.

Figure 6.6  Crush test setup example.

Typically in this test, the test sample is crushed between two flat surfaces using a force mechanism such as the hydraulic ram with a known force. The acceptable sample response in this test is for the test sample to not explode or catch fire. Figure 6.6 shows an example of a crush test setup, where the cell is placed between two flat surfaces and a force is applied using a hydraulic machine. The standards require the test samples to sustain a certain amount of crushing force without failure. The crushing force is released once the maximum is achieved in these tests. One variation of this test that is often performed is the crush to failure test. In this variation, the crushing force is applied until a cell internal short circuit is achieved. This allows for a determination of the margin of safety available for a cell or battery in an application before a crushing force leads to a failure.

### External Short Circuit Test

An external short-circuit condition can occur when a conductor bridges the output terminals of a Li-ion cell or battery. This fault condition, although usually called a short circuit in industry standards and test protocols, is technically a low resistance fault condition (typically less than 100 m$\Omega$). A short circuit condition in a fully charged multicell Li-ion battery can generate high peak currents (typically, a 2-Ah Li-ion cylindrical cell may generate peak short-circuit currents in excess of 50A). For this reason numerous industry standards detail tests that evaluate the response of a cell or battery to an external short circuit condition (typically the protocols require a fault with a resistance of less than 100 m$\Omega$). External short-circuit tests in various industry standards require the application of a direct connection between the positive and negative electrode terminals of a cell or battery to evaluate its response to a low resistance fault at its output and its ability to withstand a maximum current flow condition without causing an explosion or fire.

## References

[1] Safety Issues for Lithium-Ion Batteries, UL, https://www.ul.com/global/documents/newscience/whitepapers/firesafety/FS_Safety%20Issues%20for%20Lithium-Ion%20Batteries_10-12.pdf.

[2] Lithium Batteries: A New Approach to Risk and Safety, Intertek, http://www.intertek.com/WorkArea/DownloadAsset.aspx?id=34359752925.

[3] IEC 62281:2016, Safety of Primary and Secondary Lithium Cells and Batteries During Transport.

[4] IEC 62133: Edition 1.0, 2017, p. 19.

[5] IEC 62133-2 Edition 1.0 2017-02: Secondary Cells and Batteries Containing Alkaline or Other Non-Acid Electrolytes–Safety Requirements for

Portable Sealed Secondary Cells, and for Batteries Made from Them, for Use in Portable Applications–Part 2: Lithium systems," p. 31.

[6] Cai, W., et al., "Experimental Simulation of Internal Short-Circuit in Li-ion and Li-ion-Polymer Cells," *Journal of Power Sources*, Vol. 196, 2011.

[7] Orendoff, C., et al., "Battery Safety Testing," *Energy Storage Annual Merit Review*, Washington, DC, June 7, 2014.

[8] Lele, S., A. Arora, and K. Benson, "Predicting the Life of Li-ion Batteries Using the Arrhenius Model," *Battcon 2018*, Nashville, TN, 2018.

[9] Schmalstieg, J., et al., "From Accelerated Aging Tests to a Lifetime Prediction Model: Analyzing Lithium-Ion Batteries," *EVS27*, Barcelona, Spain, November 17–20, 2013.

[10] Smith, K., et al., "Degradation Mechanisms and Lifetime Prediction for Lithium-Ion Batteries–A Control Perspective," *2015 American Control Conference*, Chicago, July 1–3, 2015.

# 7

# Physical Construction of Battery Packs

When finalizing the design of a Li-ion battery system it is important to evaluate the physical construction and assembly of both the charger circuit and the battery protection circuit. Several critical factors such as components positioned in close proximity to the cell(s), improperly routed wires and/or current carrying tabs, excess heat application during the soldering process, and improper spot welding of tabs to the cells can all increase the risk of a battery system failure in the field. It is important to evaluate the as-built construction quality of the AC adapters, charger circuits and battery packs once prototypes have been developed as well as after engineering builds and during production.

This chapter will provide an overview of some common pitfalls that the authors have observed in the physical design and assembly of Li-ion battery systems.

## 7.1 Single-Cell Battery Packs[1]

In single-cell battery packs specifically in consumer electronic devices, it is common for the PCM to be attached directly to the cell tabs. Figure 7.1 shows an example of a single-cell battery for a consumer electronic device. In this example, the battery's PCM is installed in close proximity to the cell tabs. It is common practice in these battery packs to use Kapton tape as an insulator for the PCM. PCMs in single-cell battery packs are typically wrapped in multiple layers of Kapton tape to reduce the likelihood of damage to the cell enclosure (especially for pouch cells) due to the sharp edges of the PCM's circuit board. The installation of the PCM around the cell terminals requires special attention as inadequate insulation, improper routing of wires, or other assembly related inconsistencies can all increase the likelihood of failure.

### 7.1.1 Soldering of Cell Tabs

Although not ideal, it is not uncommon for cell tabs to be directly soldered to dedicated pads on the PCM (especially with pouch cells). This attachment process if used requires care. It is important

**Figure 7.1** Single-cell Li-ion battery pack.

---

1. The discussion in this section is also relevant to battery packs with 2 or 3 cells connected in parallel.

to ensure that the soldering process used to attach the cell tabs to the PCM is consistent and temperature controlled. The polypropylene insulation in the separator starts to melt when the cell's internal temperature approaches 175°C to 185°C, significantly reducing the strength of the separator and increasing the risk of separator failure and a cell internal short circuit [1]. For this reason, it is important to ensure that the heat applied during the soldering process is carefully controlled to prevent any damage to the cell's separator. As an example, Figure 7.2 shows the connection of the cell tabs to the PCM in a battery pack containing two cells connected in parallel. In this example, the tabs for one of the two cells connected in parallel was not properly soldered to the circuit board resulting in an increased risk of performance issues in the field. A poor cell tab to PCM connection can result in resistive heating at the connection, which can accelerate cell capacity degradation in the field and cause other issues.

Excess heat application to the cell tabs during the soldering process can also damage the cell and increase the probability of a cell failure in the field. For this reason, it is important to ensure that the soldering process is carefully controlled especially if tabs are soldered directly to a circuit board. Ideally, the cell tabs should not be exposed to the heat that may be generated during the soldering

**Figure 7.2** Inconsistent soldering of the cell tabs to the PCM can lead to both reliability and safety issues in the field.

process given the inherent risk to the cell from this process. The use of spot welding to attach the cell tabs is an alternative to the soldering process. Figure 7.3 shows an example of a single-cell battery that uses spot welding to connect the cell tabs to the PCM.

### 7.1.2 Routing of Battery Pack Wires

In consumer electronic applications, single-cell battery packs typically use wires soldered to the PCM, which are then used to connect the battery to the device that it powers (the host device). This is observed in the battery pack shown in Figure 7.1. When using wires to connect the battery pack to a host device, it is important to evaluate the routing of these wires. Care should be taken with the routing of these battery wires to ensure that the wire routing does not increase the probability of a cell/battery short-circuit condition. As an example, Figure 7.4 shows a battery pack whose wires are routed such that they are in physical contact with the cell's positive terminal tab. This wire routing scheme increases the probability of wire insulation damage and a battery short circuit due to the sharp edges of the cell tabs. This is especially true when the battery pack is exposed to mechanical shock and vibration.

Where the application requires that the battery pack wires be routed in close proximity to the cell tabs or PCM pads of opposite polarity, the use of additional insulation to isolate the wires from

**Figure 7.3** Example of a single-cell battery that uses spot welding to connect the cell tabs to the protection circuit.

**Figure 7.4** The routing of the wires in this battery pack increases the probability of a battery short-circuit condition.

the tabs/PCM pads will reduce the risk of a failure in the field. Not only should the battery pack wires be insulated from cell tabs and pads on the PCM, but the routing of the wires should also be such that the wires are not in physical contact with components on the PCM. The PCM components may operate at elevated temperatures and degrade/damage the wire insulation over a period of time. For applications where physical contact with components on the PCM is unavoidable, additional high-temperature insulation between the components and the battery pack wires reduces the likelihood of this failure in the field. The battery pack assembly and construction should reduce the likelihood of damage to the battery pack wire insulation in the field as this increases the risk of a short circuit condition.

### 7.1.3 Cell Tab Insulation

For battery packs where the PCM is directly attached to the cell tabs and is in the vicinity of the cell tabs, physical contact between the cell tabs and the cell enclosure (pouch) is inevitable. Any damage/compromise of the cell enclosure due to the sharp edges of the cell tabs increases the likelihood of a failure in the field. For pouch cells, this damage can occur when the PCM is taped (tightly) around the cell terminals to prevent it from moving. The PCM in these applications is installed on top of the cell tabs, which end up being sandwiched between the PCM and the cell pouch. In these applications, it is important to ensure that the cell tabs are adequately insulated from the cell pouch. Improper insulation of the cell tab increases the risk of damage and compromise to the cell's pouch by the sharp edges of the cell tabs. A compromise of the cell's pouch can lead to a number of issues, including cell swelling in the field.

As an example, Figure 7.5 shows a single-cell lithium polymer battery where the cell's positive tab was routed such that it was in physical contact with the cell's pouch without any insulation separating the sharp edges of the tab and the cell's pouch. This tab layout scheme increases the risk of damage to the cell pouch.

### 7.1.4 Circuit Board Insulation and Mounting

Ideally, the PCM when connected directly to the cell tabs should be mounted such that the components on the PCM do not face the cell. An improper mounting scheme increases the probability of a cell failure if a component on the PCM overheats (e.g., a resistive failure of the charge or discharge FET on the PCM). When the requirements of the application make it unavoidable for PCM components to be mounted facing the cell surface, the insulation scheme between the components and the cell enclosure should be carefully evaluated to minimize the risk of a thermal failure of a component on the PCM from damaging the cell.

Figure 7.6 shows an example of a thermal failure of a component on the PCM in a single-cell battery pack. The thermal failure

**Figure 7.5** Absence of insulation between the cell tab and the cell's pouch increases the probability of cell pouch damage in the field.

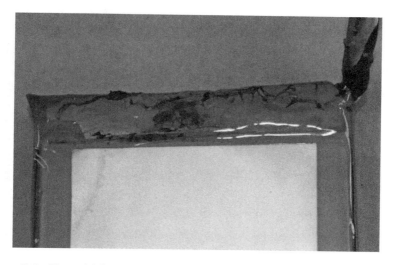

**Figure 7.6** Thermal failure of a component on the PCM in a single-cell battery pack.

of the component in this example did not lead to a thermal runaway of the cell likely due to the multiple layers of insulation that were wrapped around the PCM, which prevented the thermal failure from increasing the cell temperature to a point where the cell would have gone into thermal runaway.

### 7.1.5 Contaminants

As was discussed in Chapter 3, similar to AC adapter circuits, PCMs in battery packs are also susceptible to circuit board failures due to the presence of electrically conductive contaminants, damage during the soldering process, or improper handling. Contaminants can be especially problematic on PCMs in single-cell battery packs where the circuit board is in physical contact with the cell enclosure. Contaminant induced resistive heating on the PCM can lead to temperatures, which may lead to a cell failure. For this reason, it is important to prevent the introduction of any contaminants on the PCM during the battery assembly process.

Contaminants in the vicinity of the battery terminals can result in resistive bridging between the positive and negative terminals of the battery, which may lead to a short circuit condition. Although this condition may not necessarily result in a safety hazard in circuits that contain protection devices like fuses, it will render the battery inoperable. Figure 7.7 shows an example of a battery pack where excess contamination on the circuit board resulted in a resistive fault between adjacent terminals.

Electrically conductive and nonconductive contaminants may be present on the circuit board due to a variety of reasons. These include, but are not limited to contaminants that are inadvertently introduced due to a poorly controlled manufacturing processes, poor soldering methods, or due to environmental conditions [2]. These contaminants may be due to

- Incomplete removal of flux used during soldering;
- Dirt, dust, or oils during improper manufacturing processes;
- Improper or inadequate removal of cleaning fluids after soldering;
- Metal slivers;
- Solder bridges;
- Contaminated atmosphere during storage or assembly.

Electrically conductive contaminants (ionic contaminants) can become active in the presence of moisture and/or high humidity.

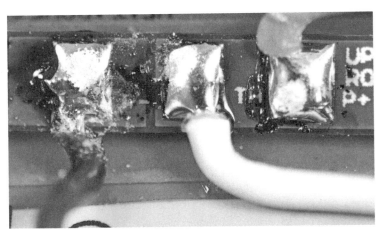

**Figure 7.7** Contaminants on the PCM increase probability of field failure.

Such contaminants, if located between copper traces or copper pads at different potentials, can create leakage current paths and cause leakage currents to travel on the surface of the PCB. Environmental contamination can result in corrosion of the copper circuit traces, damage to components, and loss of substrate electrical insulating properties. Over time, this can lead to metal migration, resulting in tin whiskers and dendrites. Whisker growth may remove material and reduce the size of the copper traces eventually resulting in an open circuit. Dendrite formation can cause conductive filaments between two traces or pads and can result in different fault conditions.

Contaminants due to environmental conditions can be present during the soldering process or may appear due to the chemical reaction of solder components. This process is an electrochemical reaction (galvanic corrosion) and an acceleration of the atmospheric corrosion occurs when two dissimilar metals are in contact with each other in the presence of an electrolyte and have a conductive path between them. Water (moisture) is an ionic liquid and is electrically conductive. This conductivity is further increased by the presence of ionic salts such as sodium chloride (NaCl).

This galvanic corrosion is likely to occur whenever two dissimilar metals are electrically connected, allowing electrons to be transferred from one to the other. In the presence of moisture, galvanic corrosion can occur and corrosion products will be produced. Further, in the presence of ionic contamination, galvanic corrosion is of specific concern for the reliability of solder connections because solder connections typically have dissimilar metals in contact with each other.

Chloride contaminants on a circuit board can lead to corrosion of the solder joints leading to poor contact issues over time. Tin-lead solders "... are very susceptible to ... corrosion" due to chlorides, carbonates, and sulfates [3]. In the presence of water, the chloride ions can form hydrochloric acid (HCl) resulting in damage to the solder connection over time [4]. Furthermore, tin-lead solder connections typically use rosin flux to ensure that the solder adheres to the copper conductors and the soldering connection is reliable. Improper removal of activated rosin flux used during the soldering process can also result in resistive faults especially in the presence of chlorides.

The risk of corrosion is not eliminated if lead-free solder is used. On the contrary, lead-free solders have an even lower resistance to corrosion than lead-tin solders. One reference states [5]

> Compared to traditional Sn-Pb (tin-lead) solders, Sn-Ag-Cu (Tin-silver-copper)[2] solders are easily corroded in corrosive environment due to their special structure. The presence of $Ag_3Sn$ in Sn-Ag-Cu solders accelerates the dissolution of tin from solder matrix into corrosive medium because of galvanic corrosion mechanism. When the corrosion is present in the solder joints, it may change the microstructure of corroded regions and decrease the mechanical properties of solder joints by providing a crack initialization.

In summary, electrically conductive contaminants can result in either an open circuit due to corrosion or can result in a conductive path between energized nodes due to, for example, whiskers. In addition, a conductive path between two traces or pads powered by low impedance sources can eventually result in a high-current fault condition. This failure may start as a dendrite or a conductive contamination between two nodes at different potentials. The relatively small leakage current (fault current) could eventually result in the carbonization of the substrate of the circuit board and create a relatively narrow carbon path. As time progresses, the carbonization continues, the fault resistance decreases and the fault current increases. Eventually the current can get high enough to cause hot spots to develop, resulting in a propagating circuit board failure with charring.

Due to the adverse consequences of contaminants on a circuit board, various techniques have been developed to make circuit boards less susceptible to contaminants. One popular technique involves conformal coating where a protective chemical coating is applied to the circuit board to protect components from both contaminants and moisture. There are many other techniques, such as potting the circuit board to insulate it from the environment.

---

2. Lead-free solder.

A discussion of the various techniques is outside the scope of this book.

## 7.2 Multicell Battery Packs

Portable consumer electronic devices equipped with multicell battery packs (e.g., laptops and tablets) typically contain six to twelve cells. Both cylindrical and pouch cells are commonly used in these battery packs. When compared to single-cell battery packs, multicell battery packs in consumer electronic devices can contain a larger number of wires/bus bars and other hardware. A typical multicell battery pack in a consumer electronic device contains the cells, a protection circuit board (BMU), voltage sense wires which connect each parallel cell group to the BMU, current carrying tabs, and a battery pack connector. Just as for single-cell battery packs, once prototypes of the multicell battery packs have been developed, the battery pack construction should be evaluated to ensure that the battery pack's construction and assembly do not elevate the risk of a field failure. The next few sections will discuss some common areas that require special attention in a multicell battery pack's construction.

### 7.2.1 Routing of Voltage Sense Wires

Pinching and stressing of wires within a battery pack can cause failures due to wear over time and potentially lead to short circuit failures within the battery pack. It is important to avoid routing wires near pinch points at different voltage potentials to minimize the risk of short circuits. In addition, it is important to ensure that wires are not routed through locations where pinching is imminent during the assembly process. When reviewing the routing of the voltage sense wires in a multicell battery pack, it is important to ensure that the wires do not run over sharp points (e.g., pins on the BMU and edges of the current carrying tabs). When routing over areas with sharp edges is unavoidable, additional insulation should be considered to minimize the risk of wire insulation damage and a potential short circuit.

### 7.2.2 Separation and Insulation of Solder Joints

In battery packs that utilize 18650 cells, it is common for a tab to be spot welded to the two ends of the cell (positive and negative) and for voltage sense wires connecting the cells to the BMU to be soldered to these tabs. These solder joints, especially when soldering is performed manually, can vary in size and shape. Poor soldering can cause formation of solder icicles that may ultimately penetrate the electrical insulation. Solder icicles have sharp points and typically appear at poorly wettable soldered joints [6]. Soldering of this nature if performed should not be done with the cells in place as this presents a risk of damage to the adjacent insulating cell sleeves, increasing the risk of a short circuit condition. Points of different potential should be separated in space to reduce the likelihood of a short circuit condition during use and abuse (e.g., drop, vibration, and shock). If separation in space is not possible given space constraints in the battery pack, additional layers of insulation should be considered to reduce the risk of a short circuit condition.

### 7.2.3 Tab Placement and Spot Welding

Tabs are often used to connect cells in parallel in multicell battery packs. It is important to ensure that the tab positioning is well controlled to reduce the risk of failures especially when the tabs are connected to the positive terminals of 18650 cells. Tabs should not be positioned such that they bypass the insulating layer installed by the cell manufacturer on the positive terminal of the cell as this can lead to a direct short circuit of the cell.

In addition to ensuring proper placement, the spot welding process used to attach the tabs to the cells should also be well controlled. A poorly controlled spot welding process can damage the enclosure of a pouch cell or lead to thermal damage in an 18650 cell. In addition, poor attachment of the cell tab can create a resistive connection between the tab and the cell potentially resulting in a hot spot. Figure 7.8 shows an example of a poorly controlled cell tab spot welding process. Excess heat application to a cell during the spot welding process can increase the probability of the cell's failure in the field. In addition, improperly placed tabs in 18650 cells may impede the operation of the cell's vent in some cases.

**Figure 7.8** The spot welding process used to attach the tab to the cell in this example appears to be uncontrolled.

## 7.3 Larger Battery Packs

Battery packs used in applications such as hoverboards, electric scooters, and power tools can contain anywhere from 20 to 100 Li-ion cells (this section does not discuss much larger battery packs such as those used in electric vehicles or grid storage applications). Similar to multicell battery packs used in portable consumer electronic devices, it is common for these comparatively larger battery packs to also contain a protection circuit (BMU) inside the overall battery pack housing. As such, most of the discussion in Section 7.2 of this chapter is relevant for these battery packs. However, these battery packs will contain a comparatively larger number of voltage sense wires, lower gauge thicker power wires, and components that may carry significantly more current than in a battery pack for a portable consumer electronic device. Hence, the requirements for multiple layers of insulation, proper wire routing, and proper separation of points at different potentials is even more important for these battery packs as a failure can result in fault currents that can be significantly higher.

### 7.3.1 Excessive Length of Voltage Sense Wires

Just as with battery packs used in portable consumer electronic devices, large battery packs also utilize voltage sense wires to communicate the voltage information for each group of cells connected in parallel to the BMU. It is important to ensure that the routing of the voltage sense wires is controlled and does not increase the risk of wire insulation damage and a short circuit. A short circuit of the voltage sense wires in these battery packs can result in a short circuit of several cells and can result in the release of a large amount of energy. Figure 7.9 shows a large battery pack where the length of the voltage sense wires connecting the cell pairs to the BMU was significantly longer than necessary and where the routing of these wires was uncontrolled. The uncontrolled wire routing coupled with the excessive lengths of the wires increases the probability of the wires coming in physical contact with the battery terminals at a different potential. Any such contact in this scenario can lead to wire insulation damage and increase the probability of a short-circuit condition.

Figure 7.10 shows another example of a battery pack with excessive wire lengths and uncontrolled wire routing. The wires in this example are routed in the vicinity of a partially insulated tab that is at a different potential than most of the voltage sense wires. Damage to the insulation of the voltage sense wires by the sharp edges of the tab can cause a short circuit of multiple cells in the battery pack.

### 7.3.2 Improper Wire Routing

Large battery packs typically utilize bus bars or low gauge wires that connect the BMU to an output connector. It is important to ensure that the routing/positioning of these bus bars and wires re-

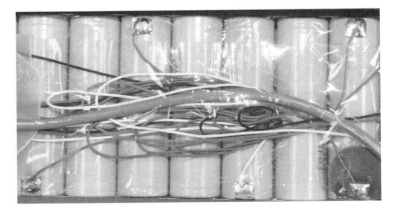

**Figure 7.9** Uncontrolled routing of voltage sense wires in a large battery pack.

**Figure 7.10** Example of uncontrolled voltage sense wire routing in a large battery pack.

duce the risk of damage to the wires under both normal and fault conditions. Physical contact between the battery pack wires and components on the protection circuit (e.g., charge and discharge FETs) can increase the probability of wire insulation damage and failure in the field.

### 7.3.3 Inadequate Insulation

Just as with multicell battery packs used in portable consumer electronic devices, it is important to ensure that wires/solder joints or tabs close to locations that are at a different potential are adequately insulated from each other to reduce the risk of a short circuit. Multiple layers of insulation are typically used in large battery packs to minimize the risk of a short-circuit condition. Figure 7.11 shows an example of a battery pack with inadequate insulation between a tab connected to a group of cells and a voltage sense wire at a different potential. The voltage sense wire in this example is adjacent to the uninsulated sharp edge of the tab and is susceptible to insulation damage. Mechanical abuse such as a high g-force due to dropping the pack, excessive vibration, or mechanical shock can cause the wire to rub against the sharp edge of the

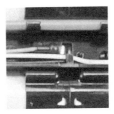

**Figure 7.11** A large battery pack with inadequate insulation between the tab connected to a cell and a voltage sense wire at a different potential.

tab. Damage to the wire insulation by the tab in this example can lead to a short-circuit condition.

### 7.3.4 Improper Torqueing of Screws

The output wires in large battery packs can sometimes be screwed into position on the BMU. In these applications, it is important to ensure that the connection of the wire to the circuit board is adequate to reduce the likelihood of a resistive fault at the connection that can initiate an overheating event. When a junction is formed by two interfacing conductors held in place by a screw, the junction has a defined contact resistance. The value of this contact resistance is primarily due to the resistivity of the two interfacing materials, the contact area between the two surfaces and the contact pressure. The formation of an oxide layer between the interfacial surfaces will result in increased contact resistance.

The contact pressure is usually selected by using a machine screw with appropriate hardware (regular washers, spring washers, and lock washers). An improperly torqued screw or a screw that is not supported by a locking mechanism (e.g., lock washers) can, over a period of time become loose, especially in the presence of mechanical shock, vibration, and thermal cycling.

A loose screw can result in a higher than normal contact resistance. The passage of current through this loose screw (with increased contact resistance) can result in increased conduction losses ($I^2R$ losses) with heating at the interfacial junctions. This poor connection with increased heating at the contact area will result in the formation of a thin film of oxides of the interfacial metals at the junction. The oxides will conduct current but with an increased contact resistance at that junction and hot spots can eventually develop that can become hot enough to glow. The rate of increase of power dissipation is dependent on several factors, such as the contact resistance, the current, and the contact pressure [7]. Heating connections have been noted at currents as low as approximately 1A [8].

Figure 7.12 shows the increase in contact temperature of a terminal block that had two of its four screws loose. In this example, it was observed that ignition occurred after an elapsed time of approximately 1,300 hours [7].

Thus, it is important to ensure that the metal screws are torqued to the required design torque and are used with the appropriate locking hardware to ensure that the manufacturer's design torque is maintained for the prescribed life of the product. Improper torque or incorrect or missing locking hardware may result in the increased power dissipation, overheating, and premature failure.

### 7.3.5 Cell Separation

For larger battery packs with tens of cells, cell spacing issues become more important since the thermal runaway of the entire battery pack can release a significant amount of energy. In these

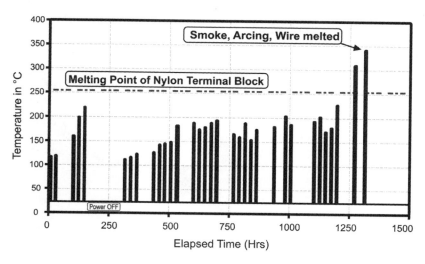

**Figure 7.12** Terminal block temperature test, temperature of pole 2, two screws loose.

battery packs, it is important to evaluate whether the thermal runaway of a single cell in the battery pack can propagate and cause all the cells in the battery pack to go into thermal runaway. Physical separation between cells (i.e., gaps that ensure that the cells do not touch each other), the use of coolants in the battery pack, or other insulation media are often used to reduce the likelihood of a single-cell thermal runaway from propagating and causing other cells in the battery pack to also go into thermal runaway.

# References

[1] Arora, A., N. K. Medora, T. Livernois, and J. Swart, "Safety of Lithium-Ion Batteries for Hybrid Electric Vehicles," *Electric and Hybrid Vehicles, Power Sources, Models, Sustainability, Infrastructure and the Market*, New York: Elsevier, 2010, Chapter 18, p. 468.

[2] Blanchard, R., N. K. Medora, et al., "Failure Analysis of Printed Wiring Assemblies," *Electronic Failure Analysis Handbook*, New York: McGraw Hill Publishing Company, 1999, Chapter 14, pp. 14.1–14.27.

[3] Vettraino, L. G., "Solder Joints," *Electronic Failure Analysis Handbook*, New York: McGraw Hill Publishing Company, 1999, Chapter 13, p. 13.41.

[4] Manko, H., *Solders and Soldering*, Fourth Edition, New York: McGraw-Hill, 2001, pp. 16–17.

[5] Song, F., S. Lee, "Corrosion of Sn-Ag-Cu Lead-Free Solders and the Corresponding Effects on Board Level Solder Joint Reliability," *Electronic Components and Technology Conference*, IEEE, 2006.

[6] http://www.technolab.de/_en/solderdict/smdhmd/soldericicle.php.

[7] Medora, N. K., "Connection Technology," *Electronic Failure Analysis Handbook*, New York: McGraw Hill Publishing Company, 1999, Chapter 17, pp. 17.1–17.69.

[8] NFPA 921, "Guide for Fire and Explosion Investigations," National Fire Protection Association (NFPA). Approved as an American National Standard (ANSI), 2011.

# 8

# Field Failures and Investigation Tools

As discussed in Chapter 5, there can be numerous causes of a Li-ion battery's failure ranging from an inadequate protection circuit, the use of an improper charge profile to charge the battery to a cell manufacturing defect. The consequence of a battery failure can range from a battery that stops functioning (i.e., cannot be charged or discharged), a battery that experiences accelerated capacity degradation, to in the worst case a battery that goes into thermal runaway. A systematic approach is necessary when investigating and determining the root cause of a Li-ion battery's failure in the field. Numerous tools and investigation techniques are available that can assist in the determination of the root cause of failure. This chapter will provide an overview of some of these tools and also review established standards and guidelines that can be used when investigating fires that are often attributed to a Li-ion battery failure in the field.

## 8.1 The Scientific Method for Investigating Battery Failures

A thermal runaway of a Li-ion battery is often cited as the cause of a fire if the area deemed to be the origin of the fire contained Li-ion batteries at the time of the fire. The condition of the Li-ion cells post the fire is often used as support for conclusions reached about the cause of the fire. However, before any conclusions can be reached about the cause of a fire, a systematic analysis should be performed to determine the true root cause of the fire. This involves an analysis of all evidence. NFPA 921 [1], the guide for fire and explosion investigations, states that "the scientific method is a principle of inquiry that forms the basis for legitimate scientific and engineering processes, including fire incident investigations.' The scientific method must be used when analyzing fires even if a failure of a Li-ion battery is suspected as the cause of the fire.

### 8.1.1 The Scientific Method

The scientific method should be used when investigating the field failure of a Li-ion battery. The scientific method is described in many different references. NFPA 921 states the following regarding the scientific method [2]:

> The scientific method is a principle of inquiry that forms the basis for legitimate scientific and engineering processes.

The scientific methodology of NFPA 921 is also described in other references. As an example, Figure 8.1 adapted from Chapter 17, Connection Technology, *Electronic Failure Analysis Handbook* [4]

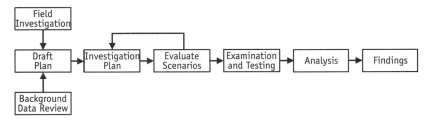

**Figure 8.1** Typical flowchart of the various steps in a failure investigation. (Adapted from [4].)

depicts the typical flowchart of the various steps in a root cause analysis that follows the scientific method.

It is important when determining the root cause of failure of the battery to understand the operating conditions of the battery leading up to the incident, analyze any historical operating data for the battery, and perform a thorough investigation of the remaining evidence to come to a root cause of the failure. Often, the damage sustained by a battery when a cell goes into thermal runaway is such that very little can be gained from analyzing the remaining evidence. In such scenarios, it is important to analyze the design of the battery charging and protection circuitry, review the construction of the cells to evaluate the overall quality of construction, and also test the response of the cell to different abuse conditions. It is not uncommon in such instances to have more than one potential root cause of failure that cannot be eliminated given the available information.

To determine the cause of a fire the following must be identified and/or ascertained [1]:

- Source of heat;
- Fuel source;
- A path of heat transfer between the heat source and the first ignited fuel.

### 8.1.2 Applying the Scientific Method to Battery Failure Investigations

Figure 8.1 shows the steps involved when using the scientific method in a failure investigation. These steps should be used when investigating the root cause of a battery failure in the field. As an example, t8he scientific method will be used as follows when determining the root cause of a Li-ion pouch cell's swelling in the field in a consumer electronics device.

- *Define the problem:* Li-ion cells are swelling in devices in the field and causing physical damage to the device enclosure.
- *Collect data:* The following is a nonexhaustive list of some of the data that should be collected when trying to determine the root cause of cell swelling in the field:

- Is the cell swelling observed in only a particular batch of cells or is the problem across the entire product space?
- What is the age of the cells when they start to swell?
- What are the typical operating and storage conditions for the cells?
- Are the charger or battery protection circuits experiencing a failure that is causing the cells to operate outside their specifications and resulting in the swelling?
- Are the batteries stored at an elevated temperature for extended period of time?
- Can the gases generated in the cells be analyzed and does the analysis point to a particular cause of the cell swelling?
- Is the cause of the cell swelling in the field associated with a product assembly issue (e.g., incorrect insulation between any metallic parts and the cell, damage to the cell during assembly)?

- *Analyze the data:* The collected data will then need to be analyzed to determine if it points to a root cause. In particular, gas analysis from swollen cells can be a very useful tool to determine the cause of cell swelling in the field. There are many references that detail the causes of gas generation in a Li-ion cell. As an example, [5] provides the mechanism of gas generation due to electrolyte decomposition in a lithium cobalt dioxide type cell that contains a cyclic alkyl carbonate and chain alkyl carbonate solution, with $LiPF_6$ as a salt under different conditions.

- *Develop hypothesis:* Once the data has been analyzed, a hypothesis can be developed. For example, if the gas analysis indicates that the cause of the swelling could be gas generation in the cell due to overdischarge, a hypothesis will need to be developed to determine how this overdischarge will occur in the field. Is the overdischarge due to a failure of the protection circuitry, a leakage path for the cell that bypasses the protection circuitry, a contaminant introduced during the assembly process that discharges the cell, or something else?

- *Test hypothesis:* The next step in the scientific method is to test the hypothesis. The testing should be performed on a statistically significant sample size to ensure that the hypothesis is valid across the population and a true cause of the field failure. The testing will need to recreate the field conditions to identify the true root cause.

- *Select final hypothesis:* The steps above should assist in the selecting of a final hypothesis that explains the field failures.

## 8.2 Analyzing Battery Failures

The following are some of the common techniques that can be used when investigating a thermal incident related to a Li-ion battery:

- Visual inspection and disassembly of the thermally damaged devices and batteries;
    - The visual inspection and disassembly is an important step to identify any burn patterns on the device and to determine whether the cause of the thermal damage is because of the battery or an external heat source that caused the battery's failure.
- Visual inspection/optical microscopic examination of both field returned thermally damaged battery packs and exemplar battery packs;
    - A visual inspection/optical microscopic examination of the thermally damaged battery pack assists in better understanding the location of damage and potential initiating location of a fault in a cell. The examination should look for signs of damage to the cell enclosure, areas where the cell vented, any heat patterns visible on the cell, and signs of external mechanical abuse (e.g., crush, penetration);
    - The examination of the exemplar battery packs typically serves as a reference for comparison with the damaged battery packs. It may also help in identifying potential manufacturing issues that may have caused the failure.
- X-ray analysis of both the thermally damaged batteries and exemplar batteries;

- When performing a root cause investigation of a battery failure, it is often necessary to evaluate the construction quality of exemplar cells to determine whether a cell manufacturing related issue is a potential cause of the field failure. X-ray analysis provides a relatively fast and low cost method for evaluating certain large-scale characteristics especially for exemplar cells. Features such as electrode alignment, internal lead damage, and electrode damage can be quickly determined using X-rays of exemplar cells. As an example, Figure 8.2 shows an example of a Li-ion pouch cell with an anode cathode misalignment.
- Computed tomography (CT) analysis;
    - CT analysis is another technique that is useful when evaluating the construction quality of exemplar cells. It allows for an accurate assessment of a cell's construction and for the detection of moderate and large sized metallic contaminants. The CT analysis may also be useful in identifying a potential short-circuit location in the failed cell.
- Destructive physical examination of both the failed and exemplar cells;
    - The examination of failed cells may aid in the identification of the fault location within a cell that triggered the failure. As an example, Figure 8.3 shows the results of the destructive physical examination of

**Figure 8.2**  An example of a Li-ion pouch cell with an anode cathode misalignment.

**Figure 8.3** The electrode winding of a failed Li-ion cell with a potential area of origin for the failure identified during a destructive physical examination of the cell.

a failed Li-ion cell. The examination identified a portion of the copper electrode on the left side of the cell that showed signs of electrical activity and that was potentially the initiating location for the cell's failure. For exemplar cells, the destructive physical examination allows for the visualization and analysis of any anomalies detected during the CT analysis and also helps to identify the presence of small metallic/nonmetallic surface contaminants, lithium plating, the quality of the ultrasonic welds, separator alignment, damage, and so forth.

- Testing of subject and exemplar battery protection and charger circuits.
  - Depending on the damage sustained by the battery protection circuit during the failure, it may or may not be possible to characterize the operation of the battery protection circuit after the incident. Successful characterization of the circuit helps in eliminating causes such as overdischarge, overcharge, and overcurrent conditions as the root cause of failure. Even if the protection circuit board sustains significant damage during the incident, it may still be in a condition where some testing is possible after the incident. In addition, x-rays of the circuit board

provides means for identifying damaged components or circuit board faults (e.g., propagating circuit board faults). The x-rays can provide evidence of a circuit board failure that may have triggered the battery pack failure. Testing should also be performed on an exemplar battery pack to ensure that the battery protection operates as it is supposed to and that the protection settings are adequate for the application.

## 8.3 Battery Failure Root Cause Analysis: A Case Study

### 8.3.1 Background

A 1S2P Li-ion battery pack (Figure 8.4) that provided power to a portable consumer electronic device was failing in the field. While a number of cells were swelling and causing damage to the battery compartment, numerous devices also experienced a battery thermal runaway condition that resulted in thermal damage to the device enclosure and led to fire hazards.

Figure 8.5 shows one of the failed Li-ion batteries from a field returned device.

The root cause analysis to determine the cause of the failures involved

**Figure 8.4**  The two-cell battery pack used in the device.

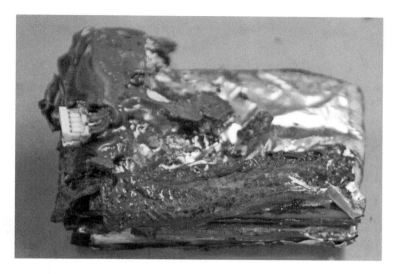

**Figure 8.5** The failed Li-ion batteries from one of the field returned devices.

- Visual inspection and disassembly of the field returned thermally damaged devices and exemplar devices;
- Design review of the devices' Li-ion battery system;
- Electrical, mechanical, and thermal testing of both subject and exemplar devices;
- Disassembly of Li-ion batteries and cells from subject devices;
- Review/analysis of the construction of exemplar Li-ion cells used in the devices;
- Accelerated aging testing of the Li-ion cells used in the devices.

### 8.3.2 Battery System Design

Figure 8.6 shows the design and various subsystems of the device's Li-ion battery system.

The various subsystems include

- AC/DC adapter that converts the AC voltage into a DC voltage that is used by the device as power for operation and to charge the Li-ion battery in the device.

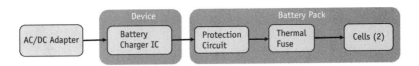

**Figure 8.6** Device battery system design.

- The battery charger circuit located in the device that converts the power from the AC/DC adapter into a voltage and current suitable for charging the Li-ion battery.
- The battery protection circuit located in the Li-ion battery that provides overcharge, overdischarge, and overcurrent protection to the two Li-ion cells in the battery.
- A thermal fuse attached to the terminals of each cell in the battery. The thermal fuse provides secondary overtemperature protection and permanently disconnects the cell it is connected to if the temperature of the fuse exceeds its trip rating.
- The Li-ion battery that contained two Li-ion cells connected in parallel.

### 8.3.3 Visual Inspection and X-Ray Analysis

A visual inspection was performed on the field returned thermally damaged devices. The inspection indicated the following:

- The area of most damage to the devices was concentrated around the battery compartment;
- At least one of the cells in the Li-ion battery in all thermally damaged devices had gone into thermal runaway and the thermal damage sustained by the devices was due to the thermal runaway of the cell(s);
- No evidence of mechanical abuse (e.g., crush, penetration, and device enclosure damage due to a drop) was found on any of the field returned devices;
- A visual inspection and review of the X-ray images of the device and battery protection circuit board eliminated a com-

ponent failure on the circuit boards as a cause of the battery failures in the field.

### 8.3.4 Battery Charger Circuit Review and Evaluation

The design of the Li-ion battery charger circuit incorporated in the devices was evaluated next to determine if the charger circuit contained all protection features commonly found in battery charger circuits designed to charge a single series battery pack. A review of the charger circuit design and testing performed on an exemplar device indicated that the charger circuit contained all protection features commonly found in battery charger circuits designed to charge a single series battery pack. Table 8.1 summarizes the protection features incorporated in the device's Li-ion battery charger circuit and provides a comparison with protection features incorporated in state-of-the-art Li-ion battery charger circuits.

The battery charger circuits in the field returned devices were tested to determine if they were functional and whether they were operating as designed. Testing was performed using exemplar Li-ion batteries to characterize the operation of the battery charger circuits in the devices. Testing indicated that the battery charger circuits in all field returned devices were operating as designed.

### 8.3.5 Battery Protection Circuit Review and Evaluation

The design of the protection circuit incorporated in the Li-ion batteries was also evaluated to determine if the circuit contained all protection features typically found in single series Li-ion battery packs. A review of the battery protection circuit design and testing performed on exemplar batteries indicated that the battery protection circuit contains all protection features commonly found in single series Li-ion battery packs. Table 8.2 summarizes the protection features incorporated in the device's Li-ion battery protection circuit and provides a comparison with protection features commonly incorporated in single series Li-ion battery packs.

Although the PCM from the subject battery packs could not be tested due to the damage they sustained during the failure, testing of the exemplar PCMs indicated that the PCMs were designed with multiple layers of protection to prevent the cells from

**Table 8.1**
Comparison between State-of-the-Art Li-ion Battery Charger Circuits and the Protection Features in the Evaluated Device's Li-ion Battery Charger Circuit

| Li-ion Battery Charger Circuits—Common Features | Device Li-ion Battery Charger Circuit |
|---|---|
| Provides a constant current-constant voltage profile (CC-CV) to charge the cells. | The charger circuit uses a CC-CV profile to charge the battery. |
| Has the ability to timeout if unable to charge cells within a specific time period. | The charger circuit times out if unable to charge the battery within approximately 12 hours. |
| Preconditions overdischarged cells by charging them at a lower charge current (i.e., the precharge current). | The charger circuit charges the battery in three phases: preconditioning, constant current, and constant voltage with a termination current of 100 mA. |
| Has the ability to monitor the cell/battery temperature and only charge the battery when the temperature is within a preset value (typically between 0°C and 45°C for charge). | The charger circuit only charges the battery for ambient temperatures from approximately 0°C to approximately 45°C. |
| Has the ability to timeout if unable to transition from precharge to fast charge mode within a specific time period. | The charger circuit times out if unable to charge the cells and transition to the fast charge mode. |
| Has the ability to detect an output short circuit condition. | The charger circuit shuts its output down on detecting an output short circuit condition. |
| Has the ability to handle an input overvoltage condition. | The charger circuit continues functioning normally for input voltages as high as 30V. |
| Terminates the charge current on detecting a communication failure with the battery pack. | The charger circuit terminates the charge current upon disconnection of the thermistor line between the charger and the battery. |

operating outside their rated specifications. This testing in addition to the x-ray analysis of the remains of the PCM from the subject batteries indicated that a failure of the PCM was an unlikely cause of the field failures.

### 8.3.6 Likely Cause of Failure

The visual inspection and electrical testing ruled out a charging circuit or battery protection circuit failure as the cause of the field failures. As was discussed in Chapter 5, the Li-ion cell has a well-defined range of operating conditions for voltage, current, and temperature. Failures generally occur when a cell is operated

**Table 8.2**
Comparison Between Features Commonly Found in Single Series Li-ion Battery Packs and the Protection Features in the Field Returned Device's Li-ion Battery Protection Circuits

| Li-ion Battery Protection Circuits—Common Features | Device Li-ion Battery Protection Circuit |
|---|---|
| Provides overcharge protection at the cell level. | The protection circuit prevents the cells from being charged above approximately 4.30V. |
| Provides two levels of output short circuit protection. | The protection circuit terminates the short circuit current in less than 50 ms. A thermal fuse is attached to each cell for secondary protection |
| Provides temperature information to the battery charger circuit . | The protection circuit provides temperature information to the battery charger circuit |
| Provides overdischarge protection at the cell level. | The protection circuit prevents the cells from being discharged below approximately 2.8V. |
| Provides two levels of over current protection during both charge and discharge. | The protection circuit provides overcurrent protection during charge and discharge. Additionally, a thermal fuse is attached to each cell for secondary overcurrent protection. |
| Generally contains an overcurrent protection device (such as a PTC, bimetal switch, or a thermal fuse) in physical contact with the cells. | A thermal fuse is attached to each cell in the battery. |

outside its specifications or exposed to mechanical, electrical, or thermal abuse conditions, such as excessive charge/discharge currents, over/under charge conditions, and over/under temperature conditions. Defects introduced in the cell due to improper or poorly controlled cell manufacturing techniques can also cause cell failures. A cell failure could lead to a cell that

- Cannot be charged and/or discharged;
- Leaks electrolyte;
- Vents and releases gases;
- Overheats;
- Experiences thermal runaway.

Table 8.3 summarizes the results of the analysis performed to determine the cause of the cell thermal runaway in the field-returned devices. Given the available information, a cell manufacturing defect was identified as a likely cause of the thermal incidents in the field.

**Table 8.3**
Root Cause Analysis of Cell Thermal Runaway in Subject Devices

| Cause | Observation/Analysis |
| --- | --- |
| External short circuit | An external short circuit of the Li-ion battery was eliminated as a cause of the field failures because the battery was designed with multiple levels of overcurrent protection for the cells, which will prevent the cells from being exposed to an external short circuit condition. In addition, the subject devices were functional after the incident and no failure on the device's circuit boards was identified, which would have resulted in a short circuit of the installed batteries. |
| Charging algorithm | A failure of the charger circuit or an improper battery charging algorithm was eliminated as a cause of the field failures because testing performed on the field-returned devices indicated that the battery charger circuit in the devices was operating as designed and contained all protection features typically found in battery charger circuits designed to charge a single-series battery pack. |
| Overcharge | A cell overcharge condition was eliminated as a cause of the field failures because testing performed on the field returned devices indicated that the battery charger circuit in the devices was operating as designed and provided the right voltage and current to charge the batteries (the charger circuit in the field-returned devices would not have overcharged the batteries). |
| Overdischarge | A cell overdischarge condition was eliminated as a cause of the field failures because testing performed on the subject devices indicated that the devices were functional and would stop discharging the battery when the battery voltage dropped below 3.2V. The batteries also contained a protection circuit that would provide secondary protection and terminate the discharge current below 2.8V. |
| Charging outside rated temperature | Charging of the Li-ion battery outside its rated temperature was eliminated as a cause of the field failures because the battery charger circuit in all field-returned devices was functional and would have prevented the cell from being charged outside its specifications. |
| Mechanical abuse | Mechanical abuse (e.g., penetration, crush) of the battery was eliminated as a cause of the field failures as no evidence of mechanical abuse was observed on either the field-returned devices or the batteries installed in the devices. |
| Cell manufacturing defect | The cells in the subject devices had sustained significant damage during the incident which made it impossible to determine if a cell manufacturing defect was a cause of the battery's thermal runaway. However, a review of exemplar Li-ion cells used in the devices indicated that the manufacturing process for the cells was not controlled. Several cell manufacturing related issues were identified during the review of exemplar cells. |

### 8.3.7 Cell Construction Review

The goal of the cell construction review was to evaluate the construction quality of the Li-ion cells. Exemplar cells from a number of date codes were inspected as part of this review. The inspection of exemplar cells indicated that the manufacturing process for the cells was not controlled. The construction of the cells was observed to be inconsistent and included several construction related issues which increase the probability of a cell failure in the field. In addition, the observed issues were not isolated to cells from a particular date code.

#### 8.3.7.1 Winding Misalignments

Figure 8.7 shows the outer winding of one of the inspected exemplar cells. An inspection of exemplar cells indicated that the process used for winding the electrodes was uncontrolled. Misaligned windings were observed on several inspected cells. Misaligned electrode windings increase the probability of lithium plating in a cell, which can lead to an internal short circuit and a cell failure.

#### 8.3.7.2 Improper Cell Tab Welding

Significant inconsistency was observed in the spot welding of the cell tabs to the electrodes on the inspected cells. Damage to the electrodes due to the spot welding process was observed on several inspected cells (Figure 8.8 shows an example). This damage due to the spot welding process can introduce metallic particles within the cell, which increases the likelihood of an internal short circuit and a cell failure.

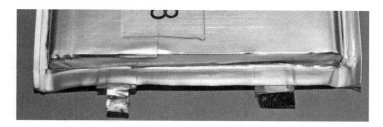

**Figure 8.7**  The process used for winding the electrodes was observed to be uncontrolled.

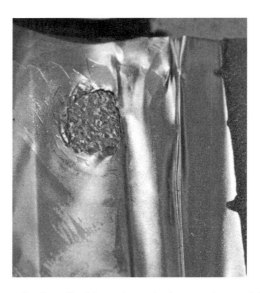

**Figure 8.8** An example of a cell with an electrode that was damaged during the cell tab spot welding process.

The process used for spot welding the tabs to the electrodes was also observed to be inconsistent on the inspected cells.

### 8.3.7.3 Damage During Manufacturing Process

Electrode damage that likely occurred during the cell assembly process was observed on several inspected cells (Figure 8.9 shows an example). Damaged electrodes increase the probability of the introduction of metallic particles in the cell. Metal particles in a cell can lead to an internal short circuit condition and cause a cell to fail.

### 8.3.7.4 Manufacturing Process Variation

The inspection of the exemplar cells indicated that the manufacturing process used for the cells was not in control. Several variations were observed in the construction of the cells including

- Different electrode lengths;
- Different spot welding processes used for the cell tabs;
- Different material used for the negative current collector.

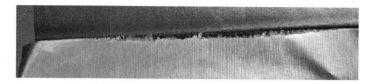

**Figure 8.9**  Damage was observed to the electrode of several inspected cells. This damage likely occurred during the cell assembly process.

#### 8.3.7.5  Inconsistent Insulation Application

The application of insulation in the cells was observed to be inconsistent on the inspected cells with insulation on tabs missing in several exemplar cells.

#### 8.3.7.6  Active Material Discoloration

Discoloration of the active material on the negative electrode was observed on some of the cells once again indicating that the manufacturing process for the cells was not fully controlled.

### 8.3.8  Summary

The review, analysis, and testing performed indicated that the most likely cause of the thermal damage sustained by the field-returned devices was due to manufacturing related issues with the device's Li-ion cells. The available evidence indicated that the manufacturing process used for the Li-ion cells was not controlled. The construction of the cells was observed to be inconsistent and included several construction related issues that significantly increase the probability of a cell failure in the field. These construction related issues were not observed only on cells from a particular batch or date code, but were observed on inspected cells from a variety of date codes.

## Appendix 8A   Investigating Failures

As discussed in Chapter 8, it is important to use the scientific method when investigating a failure and determining its root cause. NFPA 921, the guide for fire and explosion investigations states that "the scientific method is a principle of inquiry that forms the basis for legitimate scientific and engineering processes, including fire incident investigations." A number of different tools may be needed to get to the root cause of failure during the investigation. This appendix provides an overview of some of the tools that are commonly used in the process of investigating failures. The discussion in this appendix is not intended to be a comprehensive review of all tools, but rather is intended to provide the reader with an introduction to some causes of failures in electrical equipment (other than Li-ion batteries) and the investigation tools that can be used when pursuing the root cause of a fire.

### Cause of Failures in Electrical Equipment

Failures in electrical components and systems can be caused by a number of parameters and conditions. A comprehensive review of these parameters and conditions is outside the scope of this book. Some common causes of failure include

- Electrical overstress (e.g., voltage, current, and/or power overstress);
- Component/system damage due to static electricity and/or electrostatic discharge during assembly, manufacturing, installation, or repair;
- Manufacturing process variations resulting in improper or incorrect installation;
- Inconsistent quality of components;
- Mechanical stress during operation (excess shock or vibration, drop etc.);
- Inadequate protection against contaminants (ionic or organic) or corrosion;
- Inadequate protection against environmental conditions experienced during operation (e.g., elevated temperatures, humidity/moisture, or thermal cycling conditions).

### Overview of Tools Used when Conducting an Analysis of the Cause of the Failure

There are a number of tools that can be used when analyzing the available evidence to determine the root cause of failure. Depending on the complexity of the electrical system and the components used, the investigator needs to understand the features and limitations of each tool and thus determine the tools to use during the failure.

Examples of some of these tools include (some of these tools were briefly discussed in Chapter 8)

- *Radiographic (X-ray) imaging:* X-ray imaging provides another nondestructive way of reviewing the evidence. X-ray analysis involves the use of electromagnetic radiation in the soft X-ray band to produce images of objects. The image can have a magnification of 1× or higher. Since X-rays can penetrate most items, it is not necessary to remove the upper layers of an object. X-ray analysis is extensively used prior to and during a failure analysis and is very useful in identifying and examining components inside enclosures.
- *Optical microscopy (OM):* Optical microscopy (with a magnification of 1× to 1000×) is typically one of the first analysis tools that is used after the visual inspection and x-ray of the available evidence is completed. The optical microscope provides a nondestructive method of reviewing the evidence closer.
- *Computed tomography (CT) imaging:* Just like X-ray imaging, CT imaging provides a nondestructive way of analyzing the evidence. This imaging methodology is comparatively more powerful when compared to X-ray imaging and allows for the nondestructive review of portions of the evidence that may not be readily analyzed using X-ray imaging. As an example Figure 8.10 shows a CT image of a cross section of an 18650 cell. As can be seen in the figure, CT imaging provides a nondestructive method of reviewing the internal contents of the analyzed evidence.
- *Scanning electron microscopy (SEM) analysis:* SEM analysis is performed on samples with a typical magnification range of 15× to 300,000×. As compared to optical microscopy, the SEM has higher depth of field and higher resolution. As an example, Figure 8.11 shows a SEM image of a contaminant on a circuit board.

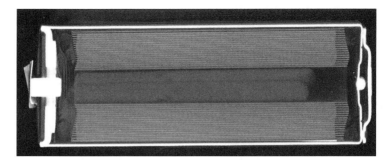

**Figure 8.10** CT image of a cross section of an 18650 Li-ion cell.

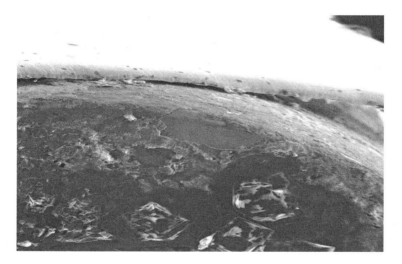

**Figure 8.11** SEM of contaminant on a circuit board.

- *Energy dispersive spectroscopy (EDS):* EDS analysis is typically combined with the SEM analysis to provide information on the elemental composition of materials observed during the SEM analysis. As an example, Figure 8.12 is an EDS spectra of a contaminant observed on a circuit board. The identification of the contaminant in this example was observed to be a chloride radical (Cl).
- *Fourier transform infrared spectroscopy (FTIR):* FTIR analysis is typically used to analyze and identify organic material. The spectrum of the material under investigation is compared

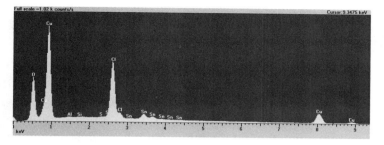

**Figure 8.12** EDS of a contaminant on a circuit board. Figure shows chloride radical (Cl).

to a library of a large number (typically 250,000 or more) of reference compounds. Figure 8.13 shows the FTIR of a contaminant (top graph) on a circuit board and comparison with other known compounds in the FTIR program library.

- *Arc mapping:* Arc mapping is a technique that uses the location of electrical arcs and arcing faults that may have oc-

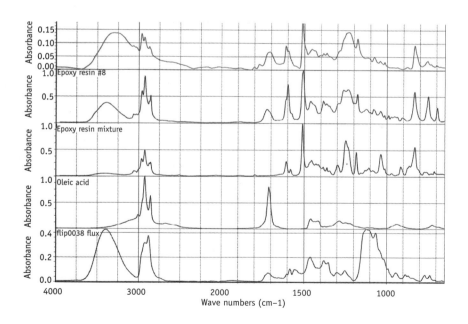

**Figure 8.13** FTIR of contaminant (top graph) on a circuit board. The contaminant signature is compared to known chemicals listed in the FTIR program library to determine if there is a good match between the subject contaminant and one or more of the known compounds in the FTIR program library.

curred during a fire to attempt to understand the progression of a fire as a function of time. NFPA 921 sec 17.4.5 [6] defines arc mapping as "...a technique in which the investigator uses the identification of arc locations or 'sites' to aid in determining the area of fire origin." Arc mapping is based on the principle that if there is a conductor that has arc marks on it, the arc which occurred furthest from the ac power source is the one that occurred the earliest. Figure 8.14 illustrates the principle of arc mapping.

In reference to Figure 8.14, arc mapping states that the arc at location C is the one that occurred earliest, since it is furthest from the ac power source and so this is the area where the fire (or hot gases created by the fire) impinged the insulated wire first and caused an electrical fault. Arc mapping assumes that once an arc occurs, the circuit breaker trips,[1] removing all power from that circuit. Thus if an arc at site B or A occurred first, then the current flow is interrupted with no further current beyond that point. Arc mapping is not an exact science, and its results are strongly dependent on certain assumptions. It can also give an incorrect conclusion for the cause of an ignition.

- *Circuit analysis using computer simulation programs such as PSpice:* Electrical circuit analysis programs such as PSpice permit the user to define the electrical parameters and variables in a system and determine the voltages and currents at the various nodes under normal operating conditions and

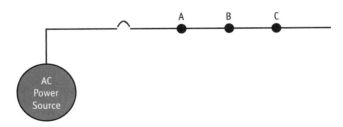

**Figure 8.14** Illustration for arc mapping.

---

1. Arcing typically is of an intermittent nature and consequently the circuit breaker may not trip.

fault conditions. The magnitude and duration of the computed variables can help in assisting in determining the cause of the failure. Modeling and simulation can also be used to vary critical variables to determine the effect of the variation of a component parameter (resistance, temperature, etc.) beyond a certain value.

# References

[1] NFPA 921, "Guide for Fire and Explosion Investigations," National Fire Protection Association (NFPA), 2011. Approved as an American National Standard (ANSI), 2017, Section 9.9.1.3..

[2] NFPA 921, Section 3.3.160.

[3] NFPA 921, Figure 4.3.

[4] Medora, N. K., "Connection Technology," in *Electronic Failure Analysis Handbook*, New York: McGraw Hill Publishing Company, 1999, Chapter 17, Figure 17.18.

[5] Kumai, et. al., "Gas Generation Mechanism Due to Electrolyte Decomposition in Commercial Lithium-Ion Cell," *Journal of Power Sources*, Vol. 81–82, 1999, pp. 715–719.

[6] NFPA 921, Section 17.4.5.

# 9
# Checklists

A significant part of the design process for a Li-ion battery system is to ensure that the battery system consists of all the necessary safety features, has all the relevant certifications, and meets all the physical packaging requirements. This chapter will provide some checklists that can be used to evaluate the components of the Li-ion battery systems that have been discussed in the various chapters of this book. These checklists are not all-inclusive and additional items may need to be added for specific applications. The aim of these checklists is to provide a general guideline that can be used by designers or at manufacturing facilities for testing and as checkpoints. The checklists do not provide a pass/fail criteria, but can be used to gather information about the battery system and make decisions on any additional work that may be required before the battery system is introduced to the field. The checklists may also be used to perform comparative evaluation between different battery and charger designs.

## 9.1 Charger Checklist

As discussed in Chapter 4, a smart charger in a portable consumer electronics device should contain a number of features. Items to evaluate when reviewing the design of the charger circuit are detailed in Tables 9.1 and 9.2.

## 9.2 Battery Checklist

Tables 9.3 and 9.4 provide a list of some items that should be reviewed when evaluating a single or multicell Li-on battery for use in a portable consumer electronics device.

**Table 9.1**
Evaluating the Battery Charger for a Single-Cell Li-Ion Battery

|  |  | Charger #1 | Charger #2 | Charger #3 |
|---|---|---|---|---|
| Visual inspection | Solder quality specifically in vicinity of battery connector appropriate? | | | |
| | Signs of excessive flux residue on circuit board? | | | |
| | Positive and negative terminal as far apart as possible on battery connector? | | | |
| | Thermistor/temperature sensor as close as possible to battery? | | | |
| Charger output | Open circuit voltage (V) | | | |
| Precharge | Battery voltage at which charger transitions to precharge mode? | | | |
| | Magnitude of precharge current (A) | | | |
| | Precharge timer setting (min) | | | |
| | Voltage at which charger transitions from precharge to fast charge mode (V) | | | |
| Fast charge | Magnitude of fast charge current (A) | | | |
| | Battery voltage at which charger transitions from constant current mode to constant voltage mode during charge cycle (V) | | | |
| | Current at which charging is terminated (A) | | | |
| | Does charger provide constant current constant voltage charge profile? (Y/N) | | | |
| | Fast charge timer settings (min) | | | |
| Temperature | Charger cutoff Temperature (Low Temperature) (°C) | | | |
| | Charger cutoff Temperature (High Temperature) (°C) | | | |

**Table 9.1** (continued)

|  |  | Charger #1 | Charger #2 | Charger #3 |
|---|---|---|---|---|
| Communications | Does charger communicate with battery? (Y/N) | | | |
| | Does charger terminate current due to lack of communications with battery (Y/N) | | | |
| | Does charger require battery ID information? (Y/N) | | | |
| | Does charger terminate current on losing battery ID info in middle of charge? (Y/N) | | | |
| | Does charger charge battery with incorrect ID or no ID? (Y/N) | | | |
| Short circuit | Does charger shut current on detecting short circuit at output? (Y/N) | | | |
| Zero volt charge | Does charger provide current to charge battery where one or more parallel cell combinations are at voltages that are less than 2V? | | | |
| Input overvoltage | Maximum input voltage at which charger continues to charge battery (V) | | | |

**Table 9.2**
Evaluating the Battery Charger for a Multicell Li-Ion Battery

|  |  | Charger #1 | Charger #2 | Charger #3 |
|---|---|---|---|---|
| Visual inspection | Solder quality specifically in vicinity of battery connector appropriate? | | | |
| | Signs of excessive flux residue on circuit board? | | | |
| | Positive and negative terminal as far apart as possible on battery connector? | | | |
| | Thermistor/temperature sensor as close as possible to battery? | | | |
| Charger output | Open circuit voltage (V) | | | |
| Precharge | Voltage of lowest individual parallel cell combination at which charger transitions to precharge mode? | | | |
| | Magnitude of precharge current (A) | | | |
| | Precharge timer setting (mins) | | | |
| | Voltage of peak individual parallel cell combination at which charger transitions from precharge to fast charge mode (V) | | | |

**Table 9.2** (continued)

| | | Charger #1 | Charger #2 | Charger #3 |
|---|---|---|---|---|
| Fast charge | Magnitude of fast charge current (A) | | | |
| | Voltage of peak individual parallel cell combination and total battery voltage at which charger transitions from constant current mode to constant voltage mode during charge cycle (V) | | | |
| | Current at which charging is terminated (A) | | | |
| | Does charger provide constant current constant voltage charge profile? (Y/N) | | | |
| | Fast charge timer settings (mins) | | | |
| Temperature[1] | Charger cutoff temperature (tow temperature) (°C) | | | |
| | Charger cutoff temperature (high temperature) (°C) | | | |
| Communications | Does charger communicate with battery? (Y/N) | | | |
| | Does charger terminate current due to lack of communications with battery (Y/N) | | | |
| | Does charger require battery ID information? (Y/N) | | | |
| | Does charger terminate current on losing battery ID info in middle of charge? (Y/N) | | | |
| | Does charger charge battery with incorrect ID or no ID? (Y/N) | | | |
| Short circuit | Does charger shut current on detecting short circuit at output? (Y/N) | | | |
| Zero volt charge | Does charger provide current to charge battery where one or more parallel cell combinations are at voltages that are less than 2V? | | | |
| Input overvoltage | Maximum input voltage at which charger continues to charge battery (V) | | | |

1. Only if charger contains thermistor that monitors battery pack ambient temperature.

## Table 9.3
### Evaluating a Single-Cell Battery

| | | Battery #1 | Battery #2 | Battery #3 |
|---|---|---|---|---|
| Visual inspection | Solder quality specifically in vicinity of battery connector appropriate? | | | |
| | Signs of excessive flux residue on circuit board? | | | |
| | Positive and negative terminal as far apart as possible on battery connector? | | | |
| | Tabs connecting cell terminals to PCM adequately insulated? (Y/N) | | | |
| | Is PCM adequately insulated from cell enclosure? (Y/N) | | | |
| | Thermistor/temperature sensor as close as possible to cell enclosure? | | | |
| Markings | Manufacturer name on battery enclosure? | | | |
| | Part number, model number and equivalent designation listed on battery enclosure? (Y/N) | | | |
| | Electrical ratings of voltage and capacity listed on battery enclosure? (Y/N) | | | |
| | List of industry standards complied with listed on battery enclosure? (Y/N) | | | |
| | Battery type listed on battery enclosure? (Y/N) | | | |
| | Date of manufacture or S/N that indicates date of manufacturing listed on battery enclosure? (Y/N) | | | |
| Compliance | Does battery need CTIA certification? (Y/N) | | | |
| | Does battery have CTIA certification? (Y/N) | | | |
| | Is battery UL listed and certified to UL 1642? (Y/N) | | | |
| | Is battery UL listed and certified to UL 2054? (Y/N) | | | |
| | Does battery have UN Testing certification? (Y/N) | | | |
| | Is battery to be sold in Europe? (Y/N) | | | |
| | Does battery have IEC certification? (Y/N) | | | |
| | Does battery have any other certification? (List) | | | |

**Table 9.3** (continued)

| | | Battery #1 | Battery #2 | Battery #3 |
|---|---|---|---|---|
| Battery design | Is a PCM attached to cell terminals? | | | |
| | Does PCM provide protection against overvoltage? | | | |
| | Does PCM provide protection against undervoltage? | | | |
| | Does PCM provide protection against overcurrent during charge? | | | |
| | Does PCM provide protection against overcurrent during discharge? | | | |
| | Does PCM provide protection against output short-circuit conditions? | | | |
| | Does battery include secondary protection through a passive device such as a PTC thermistor, a thermal fuse, or a bimetal switch? (Y/N) | | | |
| | Is passive protection device in physical contact with cell enclosure? | | | |
| | Are components on the PCM rated adequately for worst case current and voltage? (Y/N) | | | |
| | Does PCM circuit design include single point of failure which can bypass the protection circuit? (Y/N) | | | |
| | Does the drop test result in damage to the cell? | | | |
| | Does the ball drop test result in damage to the cell? | | | |
| PCM protection settings | Overvoltage protection setting (V) | | | |
| | Undervoltage protection setting (V) | | | |
| | Maximum allowable charge current (A) | | | |
| | Maximum allowable discharge current (A) | | | |
| | Peak short-circuit current (A) | | | |
| | Time to terminate short-circuit current (ms) | | | |

**Table 9.4**
Evaluating a Multicell Battery

|  |  | Battery #1 | Battery #2 | Battery #3 |
|---|---|---|---|---|
| Visual inspection | Are temperature sensors in contact with and bonded to the cell sleeves using epoxy? | | | |
| | Positive and negative terminal as far apart as possible on battery connector? | | | |
| | Are voltage sense wires routed and held in place preventing movement inside the battery pack under mechanical abuse conditions? | | | |
| | Is the separation between the voltage sense wires at their point of connection on the PCM circuit board as large as possible? | | | |
| | Are the current limiting resistors connected to the voltage sense wires on the PCM circuit board as close to the connection point on the circuit board as possible? | | | |
| | Do two levels of insulation exist between bus bars and cell cans when the two are at different potentials? | | | |
| | Are there any sharp edges on bus bars and if so do these have a potential of damaging the cells under mechanical abuse conditions? | | | |
| | Solder quality specifically in vicinity of battery connector appropriate? | | | |
| | Signs of excessive flux residue on circuit board? | | | |
| | Is PCM circuit board adequately insulated from cell enclosure? | | | |
| Markings | Manufacturer name on battery enclosure? | | | |
| | Part number, model number and equivalent designation listed on battery enclosure? (Y/N) | | | |
| | Electrical ratings of voltage and capacity listed on battery enclosure? (Y/N) | | | |
| | List of industry standards complied with listed on battery enclosure? (Y/N) | | | |
| | Battery type listed on battery enclosure? (Y/N) | | | |
| | Date of manufacture or S/N that indicates date of manufacturing listed on battery enclosure? (Y/N) | | | |

Table 9.4 (continued)

| | | Battery #1 | Battery #2 | Battery #3 |
|---|---|---|---|---|
| Compliance | Is battery UL listed and certified to UL 2054? (Y/N) | | | |
| | Is battery UL listed and certified to UL 1642? (Y/N) | | | |
| | Does battery have UN Testing certification? (Y/N) | | | |
| | Is battery to be sold in Europe? (Y/N) | | | |
| | Does battery have IEC certification? (Y/N) | | | |
| | Does battery have any other certification? (List) | | | |
| Battery design | Does PCM provide protection against overvoltage? | | | |
| | Does PCM provide protection against undervoltage? | | | |
| | Does PCM provide protection against overcurrent during charge? | | | |
| | Does PCM provide protection against overcurrent during discharge? | | | |
| | Does PCM contain secondary overcharge protection? If so, does PCM prevent cell overcharge above 4.50V in the event of a primary overcharge protection circuit failure? | | | |
| | Do cells go into thermal runaway in the event of a failure of both the primary and secondary overvoltage protection circuits? | | | |
| | Does PCM provide protection against output short-circuit conditions? | | | |
| | Does PCM contain secondary overcurrent protection? If so, does the secondary overcurrent protection circuit provide adequate protection against output short-circuit conditions? | | | |
| | Does PCM prevent the cells from being discharged at temperatures outside their ratings? | | | |
| | Does PCM prevent the cells from being charged at temperatures outside their ratings? | | | |
| | Do cells include passive protection devices, such as PTC thermistor, a thermal fuse or a bimetal switch? (Y/N) | | | |
| | Is passive protection device in physical contact with cell enclosure? | | | |

**Table 9.4** (continued)

| | | Battery #1 | Battery #2 | Battery #3 |
|---|---|---|---|---|
| Battery design | Are components on the PCM rated adequately for worst case current and voltage? (Y/N) | | | |
| | Does PCM circuit design include single point of failure which can bypass the protection circuit? (Y/N) | | | |
| | Does the drop test result in damage to the cell? | | | |
| | Does the ball drop test result in damage to the cell? | | | |
| | Does PCM disable battery pack if cell imbalance exceeds 0.5V? | | | |
| | Does PCM disable pack under voltage sense line(s) failure condition? | | | |
| PCM protection settings | Primary overvoltage protection setting (V) | | | |
| | Primary undervoltage protection setting (V) | | | |
| | Secondary overvoltage protection setting (V) | | | |
| | Secondary undervoltage protection setting (V) | | | |
| | Maximum allowable charge current (A) by primary protection circuit | | | |
| | Maximum allowable discharge current (A) by primary protection circuit | | | |
| | Maximum allowable charge current (A) by secondary protection circuit | | | |
| | Maximum allowable discharge current (A) by secondary protection circuit | | | |
| | Peak short-circuit current (A) | | | |
| | Time to terminate short circuit current (ms) | | | |
| | Maximum allowable cell imbalance voltage (mV) | | | |
| | Time to disable battery pack under voltage sense line(s) failure condition | | | |

# Glossary

This chapter defines some commonly used terminology when discussing Li-ion battery systems. These terms are used frequently throughout the book. The definitions given in this chapter have been obtained from several different references.

**Abuse**  Use of a product in a manner that is not intended by the manufacturer but exposes the product to extreme conditions or environments [1].

**Active material**  The cell constituent that participates in the charge/discharge reaction.

**Adapter**  A device that transforms the available power from an external source (AC wall outlet, automobile outlet, etc.) to the power used by the host.

**Aggregate lithium content**  The sum of the grams of lithium content contained by cells comprising a battery [2].

**Aging**  Refers to the changes that take place within a cell, such as loss of capacity as the cell ages. The effects of aging can be accelerated by temperature or improper use of the cell (elevated charge, discharge currents, etc.).

**Ambient temperature** The average temperature of the surrounding air that comes in contact with the battery. (e.g., the air temperature inside the device in which the battery is installed.)

**Ampere hour (Ah)** Unit commonly used for a battery's capacity. An ampere hour is a unit of charge in the battery that will allow one ampere of current to flow for 1 hour.

**Anode** An electrode through which current enters any conductor of the nonmetallic class. Specifically, an electrolytic anode is an electrode at which negative ions are discharged, or positive ions are formed, or at which other oxidizing reactions occur [12].

**Battery** Two or more cells connected together electrically. Cells may be connected in series, parallel, or both to provide the required operating voltage and current. It is also common to use this term for a single cell along with its protection circuitry. This term is also often used in the context of a single cell.

**Battery (energy) efficiency** The energy efficiency, expressed as a percentage of the ratio of the Watt-hour output of the battery, to the Watt-hour input required to restore the initial state of the charge.

**Battery nominal voltage** The nominal voltage of one cell multiplied by the number of series-connected cell groups in the battery.

**Battery system** The combination of the cell, battery pack, host device, and power supply or adapter.

**C-rate (specific current)** This term is used to express the charge and discharge rates. The symbol C is followed by a number, or preceded by a number or decimal portion of an integer, or divided by a number. When used to represent the charge and/or discharge current, the current is referenced to a specific cell/battery capacity. For example, if a cell rated for 2 Ah is being discharged at 500 mA, its discharge rate is 0.25C or C/4.

**Calendar life**  Calendar life expresses the theoretical lifetime of a battery when sitting at rest at a given temperature and state of charge [5].

**Cathode**  An electrode through which current leaves any conductor of the nonmetallic class. Specifically, an electrolytic cathode is an electrode at which positive ions are discharged, or negative ions are formed, or at which other reducing reactions occur [12].

**Cell**  The basic electrochemical unit, characterized by an anode and cathode used to receive, store, and deliver electrical energy. The cell is characterized by a nominal potential that is typically 3.7V for a Li-ion cell with a cobalt dioxide positive electrode and a graphite-based negative electrode. The word battery is often used to refer to a cell although technically this is incorrect.

**Cell balancing**  In batteries with cells or cell groups connected in series, cell-balancing techniques aim to distribute energy equally among the cells in a battery pack. Without cell balancing, a portion of the battery's capacity will be wasted.

**Charge (of a cell/battery)**  The conversion of electrical energy into chemical energy within a cell.

**Charging algorithm**  The set of rules and decisions used to determine the voltages and currents applied to the cell, cells, and/or battery pack as a function of time, temperature, or other parameters [1].

**Constant current–constant voltage charge**  The charge profile typically used to charge Li-ion batteries. In this profile, the charger first operates in the constant current mode where the charge current is maintained at a particular constant value until the battery voltage reaches a certain point. The charger then transitions to the constant voltage mode where the charger output voltage is maintained at a constant value.

**Columbic efficiency** The ratio of the output of charge from a cell/battery to the input of charge to the cell/battery (i.e., the ratio of the total charge extracted from a cell/battery to the total charge inputted into a cell/battery over a full charge cycle).

**Cut-off voltage** The voltage at which a discharge of the cell/battery is terminated.

**Cycle** A sequence of fully charging and fully discharging a rechargeable cell or battery. Sometimes additional modifiers are used to describe how much of the cell's capacity was removed during the discharge (e.g., shallow cycle or deep cycle).

**Cycle life** Cycle life is related to the aging of a battery during repeated charge/discharge cycles. The cycle life is dependent on the protocol used for charging and discharging the battery pack and also depends upon the battery's condition during the rest period between the charge and discharge cycles [5].

**Depth of discharge** Measure of how deeply the battery is discharged or the ampere-hours removed from a fully charged battery. It can also be expressed as a percentage of its rated capacity at the applicable discharge rate. It is commonly referred to as DOD. As an example, if a 10-Ah battery is 100% fully charged, it means DOD of the battery is 0% or 0 Ah.

**Discharge (of a cell/battery)** The conversion of chemical energy into electrical energy within a cell and the subsequent transfer of this energy into an external load.

**Discharge rate** The rate in amperes at which current is delivered by the battery.

**Electrode**  The site, area, or location at which the electrochemical reaction takes place.

**Electrolyte**  A conducting medium in which the flow of electric current takes place by migration of ions [12].

**End of discharge voltage**  The minimum voltage at which a battery is intended to operate during a discharge event for a given application.

**End of life**  End of life is commonly defined using some measure of faded performance relative to beginning of life [5].

**Failure mode**  The way in which a system can fail. The failure can be due to a variety of reasons, such as a defect, operation outside rated specifications, and abuse conditions.

**Fault**  A condition that results in a device or its component not operating in its desired manner. Faults can include conditions such as a short circuit, open circuit, and intermittent connections.

**Foreseeable misuse**  Use of a product in a way that is not intended by the manufacturer but may result from foreseeable use in the field.

**Fully charged**  A rechargeable battery or cell that has been electrically charged to its design rated capacity [2].

**Fully discharged**  A rechargeable cell or battery that has been electrically discharged to its endpoint voltage as specified by the manufacturer. A discharge cut-off voltage of 3.0V is common for Li-ion cells with a cobalt dioxide based positive electrode and a graphite-based negative electrode.

**Host**  The device that is powered by a battery and/or charges the battery (e.g., a smartphone [1]).

**Intended use**  The use of a product in accordance with its specifications or instructions provided by the manufacturer [1].

**Internal impedance**  The resistance of a cell to an alternating current at a specific frequency.

**Internal resistance**  The DC resistance of a cell to an electric current within a cell (i.e., the sum of the ionic and electrical resistances of the cell components).

**Internal voltage drop**  The product of the current passing through the cell and its internal resistance.

**Lithium content**  The lithium content of a battery equals the sum of the grams of lithium content contained in the component cells of the fully charged battery [2]. The equivalent lithium content (ELC) for any battery is found by multiplying the ampere-hour capacity of each cell by 0.3 and then multiplying the result by the number of cells in the battery [7].

**Nominal voltage**  The approximate value of the voltage used to designate or identify a cell/battery [2].

**Open-circuit voltage**  The voltage of a cell with no current flow in either direction after the cell has had time to stabilize. It is commonly referred to as OCV.

**Overcharging (a cell)**  The forcing of current through a cell/battery after it has been fully charged.

**Overcurrent**  A condition where the charge current to a battery or a discharge current from a battery exceeds the current magnitude specified by the battery manufacturer.

**Overdischarge**  Discharge past the point where the full capacity of the battery has been obtained.

**Overvoltage** A condition where the charge voltage to a battery exceeds the voltage specified by the battery manufacturer.

**Pouch cell** A cell that is often prismatic in shape and whose contents are enclosed within a sealed flexible pouch rather than a rigid casing [4].

**Rated capacity** The capacity assigned to a cell by the cell manufacturer for a particular discharge current, temperature, and end of discharge voltage.

**Secondary cell** An electrochemical cell that is capable of being discharged and then recharged.

**Self-discharge** The performance of a cell changes due to parasitic reactions even when the battery is at rest due to the increase of the cell's internal impedance and a decay in its capacity. Capacity losses in a cell can be reversible or irreversible. The reversible capacity loss in a cell due to these parasitic reactions is known as self-discharge [9]. IEEE defines it as the process by which the available capacity of a battery is reduced by internal chemical reactions [12].

**Self-discharge rate** The amount of capacity reduction occurring per unit time in a battery as a result of self-discharge.

**Separator** An ionic, permeable, nonconductive spacer used to prevent metallic contact between electrodes of opposite polarity within a cell.

**Series** The interconnection of cells in such a manner that the positive terminal of the first cell is connected to the negative terminal of the second cell, and so on.

**Service life** The length of time in which a fully charged battery is capable of delivering at least a specified percentage of its rated capacity. For most Li-ion batteries, this percentage is either 80% or 70%.

**Shelf life**   The duration of storage under specified conditions at the end of which a cell or battery retains the ability to give a specified performance.

**Shunt**   A calibrated resistance of very low ohmic value used to measure current in a circuit. The monitoring circuit measures the voltage drop across the shunt resistance and thus determines the magnitude of the current. A shunt is inserted in series in the circuit where the current is to be measured.

**State of charge**   In general, the state-of-charge of a battery is defined as the ratio of its current capacity to its nominal capacity. The nominal capacity is typically provided by the cell/battery manufacturer and represents the maximum charge that can be stored in the cell/battery [10].

**String**   A common way to refer to a number of cells connected together in series to form a battery.

**Thermal runaway**   Self-heating that rapidly accelerates to high temperatures [11]. Occurrences of thermal runaway in Li-ion cells are exothermic and can pose a fire hazard [6].

**Thermistor**   A commonly used component for temperature sensing in Li-ion batteries.

**Trickle charge**   A charge given to a battery with no external load connected to it, to maintain it in a fully charged condition.

**Vent**   A mechanism that allows for the escape of gases from within a cell.

**Venting**   Release of excessive internal pressure from a cell/battery in a manner intended by design to preclude rupture or explosion.

**Voltage drop**   The voltage difference between the voltages measured at the cell terminals and a point downstream, such as at the battery connector. Excessive voltage drop across current carrying wires is undesirable as it results in a loss in energy in the form of heat, thus reducing performance.

# References

[1] IEEE Standard for Rechargeable Batteries for Cellular Telephones, IEEE Std. 1725-2011, June 10, 2011.

[2] *Recommendations on the Transport of Dangerous Goods: Manual of Tests and Criteria*, Fifth Revised Edition, Amendment 1, New York/Geneva: United Nations, 2009.

[3] IEC 62133, International Standard, Edition 2.0, 2012-12.

[4] UL 1642, Standard for Safety–Lithium Batteries, 2012.

[5] Santhanagopalan, S., et al., *Design and Analysis of Large Lithium-Ion Battery Systems*, Norwood, MA: Artech House, 2015, p. 84.

[6] Arora, A., N. K. Medora, T. Livernois, and J. Swart, "Safety of Lithium-Ion Batteries for Hybrid Electric Vehicles." In G. Pistoia (ed.), *Electric and Hybrid Vehicles, Power Sources, Models, Sustainability, Infrastructure and the Market*, Amsterdam: Elsevier, 2010, Chapter 18, pp. 463–491.

[7] https://en.wikipedia.org/wiki/Lithium-ion_battery.

[8] Reddy, T. B. (ed.), *Linden's Handbook of Batteries*, Fourth Edition, New York: McGraw Hill, 2011.

[9] Iglesias, E. R., et. al., "Global Model for Self-Discharge and Capacity Fade in Lithium-Ion Batteries Based on the Generalized Eyring Relationship," *IEEE Transactions on Vehicular Technology*, Vol. 67, No. 1, 2017, pp.104–113.

[10] Chang, W.Y., "The State of Charge Estimating Methods for Battery: A Review," *ISRN Applied Mathematics*, Vol. 2013, Article ID 953792, 2013, p. 7.

[11] Babrauskas, V., *Ignition Handbook*, Issaquah, WA: Society of Fire Protection Engineers, 2003, p. 20.

[12] *IEEE 100: The Authoritative Dictionary of IEEE Standard Terms*, Seventh Edition, Institute of Electrical and Electronics Engineers (IEEE), 2000.

# About the Authors

**Ashish Arora** is a principal engineer at Exponent, Inc., an engineering and scientific consulting firm. He works extensively on energy storage systems in the consumer products, aviation, automobile, and utility industries. He has spent the last 15 years performing design reviews, testing, field failure investigations, and auditing of battery manufacturers and is internationally recognized for his expertise in this field. He has consulted for a wide variety of clients that use Li-ion batteries for energy storage in their applications and has publications, presentations, book chapters, and a patent in this discipline. Ashish has a master of science degree in electrical and computer engineering from Purdue University and an M.B.A. from Indiana University.

**Sneha Lele** is a senior associate in the Electrical and Computer Engineering Practice at Exponent, Inc. At Exponent, she is extensively involved in battery-related analyses, root cause investigations, and testing of products for accelerated aging of batteries and other electronic components. Her expertise is in design and quality reviews of household and commercial products, automotive electronics, and industrial systems with a focus on system safety and reliability. Dr. Lele received her Ph.D. in electrical and computer engineering from the University of Western Ontario, Canada. Prior to joining Exponent, Dr. Lele also worked at Siemens (India), Advanced Micro Devices (Canada), and ADVA Optical Networking (United States).

**Noshirwan Medora** is a senior managing engineer in the Electrical Engineering and Computer Science Practice at Exponent. He has over 35 years of experience in the areas of power electronics and electrical and electronic products and has worked on lead-acid, NiMH, NiCd, and lithium-ion batteries, AC/DC motors and motor drives, transformers, circuit breakers, inverters, converters, uninterruptible power supplies (UPS), regulators, SMPS (switch mode power supplies), AC/DC high voltage systems, smartmeters, and appliances. He has particular expertise in automotive electronics, utility power systems, transportation systems, and industrial electronics. Mr. Medora has a master of science degree in electrical engineering and an engineer's degree in electrical engineering from the Massachusetts Institute of Technology.

**Shukri Souri** is a principal and a corporate vice president at Exponent. He is based in, and director of, Exponent's New York office, and is also the director of Exponent's Electrical Engineering and Computer Sciences Practice. His expertise is in microelectronics and computing systems and his professional activities involve advising industrial and legal clients as well as government entities on science and technology matters addressing issues related to intellectual property, product reliability, and failure analysis. He has led complex investigations involving electronics, controls, and mobile energy storage for safety-critical applications in the medical device, automotive, aviation, and consumer electronics industries. Dr. Souri received his M.S.E.E. and Ph.D. degrees in electrical engineering from Stanford University. He also received his B.A. (Hons) in engineering science from Balliol College, Oxford.

# Index

18650, 26, 27, 30, 41, 174, 199

## A

Abuse, 7, 71, 76, 103, 106, 130, 135
AC adapter, 3, 7, 14, 18, 39
Accelerated rate calorimetry, 23
Active material, 25, 197
Aggregate lithium content, 215
Aging, 30–33, 70, 88, 112, 152, 153
Allowable charge, 95, 118
Ambient temperature, 23, 30, 33, 40, 70, 76, 78, 92, 122, 147, 150, 155
Ampere hour, 216
Anode, 17, 19, 20, 21, 23, 25, 83, 154, 186, 216
Arc mapping, 201
Arrhenius, 33, 152–155
ASTM, 57, 58, 59

## B

Battery management system, 28, 43
Battery nominal voltage, 216
BMS, 28, 42–45
BMU, 28, 42–44, 121–124, 173–176, 178
Burn hazard, 54, 57–59, 62, 76
Bus Bars, 111

## C

Cathode, 17, 19, 20–25, 83, 84, 186, 217
CC-CV, 105, 117, 192
Cell assembly, 104, 196, 197
Cell balancing, 43, 111, 112, 115, 119, 122, 217
Cell Imbalance, 95, 98, 112, 129
C-FET, 106–108
Charge circuit, 39, 40, 43, 47, 48, 83, 85, 96, 98, 108
Charge cut-off, 91, 127
Charging algorithm, 105, 194, 217

Checklists, 205
CID, 27
Clearance, 72
Columbic efficiency, 218
Computed tomography, 199
Consumer electronic devices, 13, 17, 26, 27, 42, 45, 47, 52, 62, 66, 77, 83, 101, 136, 157, 166
Contact resistance, 67, 178
Contactors, 115
Contaminants, 53, 73, 124, 169
Converters, 48, 77
C-Rate, 85
Creepage, 71, 144
Current collector, 25, 105, 196
Current limiting resistors, 123
Cut-off voltage, 218
Cycle life, 18, 23, 33
Cycleable lithium, 25
Cylindrical, 26, 27

**D**

Depth of discharge, 18, 32
D-FET, 99, 100, 106
DFMEA, 120
Discharge cut-off, 127
Discharge rate, 19

**E**

Efficiency, 38, 63
Electric shock, 52
Electric vehicles, 17, 20, 38, 43, 115, 136, 159, 175
Electrochemical stripping, 89
Electrolyte, 19–21, 25
End of discharge voltage, 219
Energy dispersive spectroscopy, 200
Energy storage system, 38
EOL, 29

**F**

Fast-charge, 85
Flux residue, 93, 124
Foreseeable misuse, 144
Fully charged, 30
Fully discharged, 86

**G**

Galvanic corrosion, 171, 172
Graphite, 21, 23, 83, 84
Grid storage, 38, 47, 175

**H**

HAZOP, 120
Hi-pot, 75
Host, 17, 21, 38, 39, 42, 90, 145, 146, 160
Hypothesis, 184

**I**

IEC 60950, 71, 138
IEC 61960, 143, 156
IEC 62133, 136, 137, 143–146, 148, 156, 157
IEEE Std. 1625, 38, 146
IEEE Std. 1725, 24, 66, 145–147
Insulation, 53, 70, 72, 75, 77, 104, 122, 144, 164–169, 173–178, 194, 211
Intended use, 220
Intercalation, 20
Internal impedance, 104
Internal resistance, 24, 125
Internal voltage drop, 220

**J**

JIS, 148, 156

## L

Lead acid, 18, 19
Leakage current, 171, 172
Lithium content, 220
Lithium plating, 88, 89, 187

## M

Markings, 117, 209, 211
MOVs, 50

## N

Nail penetration, 149
NFPA 921, 182, 197
NiCd, 5, 8, 17–19
NiMH, 19, 37
NTCs, 109

## O

Open-circuit voltage, 125
Optical microscopic examination, 185
Opto-coupler, 52
Overcharging, 88
Overcurrent, 106
Over-discharge, 40
Overstress, 198
Over-voltage, 49, 64

## P

PCM, 40, 108, 109, 111, 112, 118, 121–123, 167
PolyZen, 64
Pouch, 26–28
Precharge, 85, 89, 90, 198, 199
Primary protection, 114, 129
Prismatic, 26
Propagating PCB failure, 53
PTC, 27, 40, 48, 64, 110

## R

Rated capacity, 30
Rectifier, 51, 52, 61, 69
Root cause, 181, 182, 192, 193

## S

SAE International, 156
Safety timers, 86
Scanning electron microscopy, 199
Scientific method, 182, 183
Secondary cell, 143
SEI layer, 25, 30, 87, 88
Self-discharge, 19, 125
Separator, 20, 24, 105, 165
Service life, 37
Shelf life, 31, 147
Short-circuit, 63, 75, 78, 89, 96, 102, 104, 107, 124, 128, 146, 151, 160, 210
Shunt, 51
SMPS, 48
Solder bridges, 170
Spectroscopy, 200
Spot welding, 124, 166, 174, 195
Spring loaded connections, 68
Storage temperature, 31, 68
String, 45, 111
Surface temperature, 22, 56, 102
Switching, 51, 52, 61, 77, 99, 112

## T

Thermal characterization, 75, 112
Thermal fuse, 40, 50, 109, 190
Thermal runaway, 23, 50, 67, 89, 102, 125, 150, 169, 180
Thermal stability, 22, 150
Thermesthesiometer, 58, 59
Thresholds, 131, 132

Torqueing, 178
Transformer, 51
Trickle Charge, 93

## U

UL 1642, 137, 142, 155
UL 2054, 137, 141, 155
UL 60950-1, 56, 72, 116, 121
UL 62133, 143
UL1642, 22
UN transportation, 136, 139
Useful capacity, 18

## V

Vent, 26, 70

Voltage drop, 85, 107
Voltage sense wires, 42, 111, 123, 173

## W

Winding, 89, 104, 195
Wireless chargers, 98

## X

X-ray, 185, 190, 199

## Y

yPxS, 111

# Recent Artech House Titles in Power Engineering

Andres Carvallo, Series Editor

*Advanced Technology for Smart Buildings,* James Sinopoli

*The Advanced Smart Grid: Edge Power Driving Sustainability, Second Edition,* Andres Carvallo and John Cooper

*Battery Management Systems, Volume I: Battery Modelings,* Gregory L. Plett

*Battery Management Systems for Large Lithium Ion Battery Packs,* Davide Andrea

*Battery Power Management for Portable Devices,* Yevgen Barsukov and Jinrong Qian

*Big Data Analytics for Connected Vehicles and Smart Cities,* Bob McQueen

*IEC 61850 Demystified,* Herbert Falk

*Design and Analysis of Large Lithium-Ion Battery Systems,* Shriram Santhanagopalan, Kandler Smith, Jeremy Neubauer, Gi-Heon Kim, Matthew Keyser, and Ahmad Pesaran

*Designing Control Loops for Linear and Switching Power Supplies: A Tutorial Guide,* Christophe Basso

*Electric Power System Fundamentals,* Salvador Acha Daza

*Electric Systems Operations: Evolving to the Modern Grid,* Mani Vadari

*Energy Harvesting for Autonomous Systems,* Stephen Beeby and Neil White

*GIS for Enhanced Electric Utility Performance,* Bill Meehan

*Introduction to Power Electronics,* Paul H. Chappell

*Introduction to Power Utility Communications,* Dr. Harvey Lehpamer

*IoT Technical Challenges and Solutions,* Arpan Pal and Balamuralidhar Purushothaman

*Lithium-Ion Battery Failures in Consumer Electronics,* Ashish Arora, Sneha Arun Lele, Noshirwan Medora, and Shukri Souri

*Microgrid Design and Operation: Toward Smart Energy in Cities,* Federico Delfino, Renato Procopio, Mansueto Rossi, Stefano Bracco, Massimo Brignone, and Michela Robba

*Plug-in Electric Vehicle Grid Integration,* Islam Safak Bayram and Ali Tajer

*Power Grid Resiliency for Adverse Conditions,* Nicholas Abi-Samra

*Power Line Communications in Practice,* Xavier Carcelle

*Power System State Estimation,* Mukhtar Ahmad

*A Systems Approach to Lithium-Ion Battery Management,* Phil Weicker

*Signal Processing for RF Circuit Impairment Mitigation in Wireless Communications,* Xinping Huang, Zhiwen Zhu, and Henry Leung

*The Smart Grid as An Application Development Platform,* George Koutitas and Stan McClellan

*Smart Grid Redefined: Transformation of the Electric Utility,* Mani Vadari

*Synergies for Sustainable Energy,* Elvin Yüzügüllü

*Telecommunication Networks for the Smart Grid,* Alberto Sendin, Miguel A. Sanchez-Fornie, Iñigo Berganza, Javier Simon, and Iker Urrutia

For further information on these and other Artech House titles, including previously considered out-of-print books now available through our In-Print-Forever® (IPF®) program, contact:

Artech House
685 Canton Street
Norwood, MA 02062
Phone: 781-769-9750
Fax: 781-769-6334
e-mail: artech@artechhouse.com

Artech House
16 Sussex Street
London SW1V 4RW UK
Phone: +44 (0)20 7596-8750
Fax: +44 (0)20 7630-0166
e-mail: artech-uk@artechhouse.com

Find us on the World Wide Web at: www.artechhouse.com